Successful Proofs for Riemann hypothesis, Polignac's and Twin prime conjectures using Elementary-Emergent Fundamental Laws

Introduction incorporating Medical Materials

BOOK SERIES 1

John   Ting

DEDICATION & COPYRIGHT

To my father Samuel, my wife Jocelyn and my children Jonah, Joelle, Jethro, Jonty and Jelena. To the loving memory of my mother Grace who passed away on February 13, 2016.

Author corresponding details: Dr. John Ting, Dental and Medical Surgery, 729 Albany Creek Road, Albany Creek QLD4035, Australia. Email: jycting@hotmail.com

Cover design: July 7, 2016 picturesque sunrise photo at Brighton shoreline in Brisbane, Australia by author.

Independently published in Kindle paperback and eBook as Riemann hypothesis, Polignac's and Twin prime conjectures, Fundamental Laws, and Medicine BOOK SERIES by Kindle Direct Publishing on Amazon

Kindle paperback ISBN: 9781074451820

ACKNOWLEDGEMENTS

I am thankful to Jeanette "Jenny" O'Hagan (writer and speaker with medical degree), Rodney "Rod" Williams (civil engineer with mathematical degree) and Anthony "Tony" O'Hagan (software engineer with mathematical degree) for invaluable advices, editing and proofreading of this book.

I have my personal homepage Research by Dr. John Y C Ting on Maths, Physics, Science, Medicine, and Religion https://jycting.wordpress.com and welcome feedbacks from interested readers. Jenny can be found at Linkedin https://www.linkedin.com/in/jeanetteohaganwrites/. Rod can be found at Linkedin https://www.linkedin.com/in/rod-williams-1b88a346/. Tony from Digital Ministry https://www.crossover.org.au/digital-missionaries/ with his personal homepages Internet Missions http://internetmissions.org and Crusade for Christ https://give.cru.org/0877152 also works part time at Power to Change https://www.powertochange.org.au. For a good cause, please donate generously at Give to a Missionary https://www.powertochange.org.au/give/give-to-a-missionary by selecting O'Hagan, Tony & Jenny.

PREFACE

Open problems in Number theory of Riemann hypothesis, Polignac's and Twin prime conjectures have been unsolved for over 150 years. They are finally solved as Incompletely Predictable problems in 2019. Riemann hypothesis belongs to one of the seven Millennium Prize Problems in mathematics stated by Clay Mathematics Institute on May 24, 2000. I serendipitously cross path with the Institute's website Riemann Hypothesis

https://www.claymath.org/millennium-problems/riemann-hypothesis

when I had the honor on Thursday March 10, 2016 in permanently altering its information content from previous incorrect [sic] "...the first 10,000,000,000 solutions" to current correct [sic] "...the first 10,000,000,000,000 solutions". Configured as 'Introduction incorporating Medical Materials BOOK SERIES 1', Chapters 1 to 3 as beginning part of this book contain introductory materials with Chapter 1 explaining the important Fundamental Laws. Chapters 4 to 8 as middle part of this book concentrate on describing in layman's terms how rigorous proofs for our three open problems are derived by me (as two complete research papers first published in viXra and reproduced in Appendix 1 and 2). Chapters 9 to 10 as end part of this book contain materials that speculate on important role of Umbral ("Shadow"), Mathieu and Monstrous Moonshine in String theory potentially uniting Einstein General Relativity and Quantum gravity, and refute a common misconception that solving Riemann hypothesis will lead to E-Commerce apocalypse. My exotic A228186 Hybrid integer sequence published on August 15, 2013 in The On-line Encyclopedia of Integer Sequences (OEIS) is also fully outlined in this book.

CONTENTS

PROLOGUE

Re: http://www.claymath.org/millennium-problems/riemann-hypothesis

Naomi Kraker <admin@claymath.org>

Thu 10/03/2016 7:56 PM

To: john ting <jycting@hotmail.com>

Dear Dr. Ting,

Thank you very much for your email and for drawing our attention to the error on our website, which I have now corrected. I think the text was probably picked up from an old site. Please let us know when even more zeroes need to be added!

With best regards,

Naomi

Naomi Kraker
Administrative Manager
Office of the President, Clay Mathematics Institute
Andrew Wiles Building
Radcliffe Observatory Quarter
Woodstock Road
Oxford OX2 6GG U.K.
Email: admin@claymath.org
Tel: +44 (0)1865 615155
www.claymath.org

On 10 Mar 2016, at 07:58, john ting <jycting@hotmail.com> wrote:
Thursday 10 March 2016

Dr John Yuk Ching Ting
729 Albany Creek Road
Albany Creek
Queensland 4035
Australia
Email: jycting@hotmail.com

admin@claymath.org
Clay Mathematics Institute
70 Main St
Suite 300
Peterborough, NH 03458
USA

Dear Sir / Madam,

I refer to your website on Riemann Hypothesis "http://www.claymath.org/millennium-problems/riemann-hypothesis" last updated on Tuesday 12:34 pm 1 March 2016 - see attached document. I believe that its last paragraph "This has been checked for the first 10,000,000,000 solutions. A proof that it is true for every interesting solution would shed light on many of the mysteries surrounding the distribution of prime numbers." is incorrect in that "...the first 10,000,000,000 solutions" should be "...the first 10,000,000,000,000 solutions" instead.

Please let me know whether the above observation by me is correct or not.

Many thanks.

Kind regards,

John Yuk Ching Ting

<Riemann Hypothesis Clay Mathematics Institute Last update 01May2016 1234pm.pdf>

Here is a brief synopsis on this book. Title: *Successful Proofs for Riemann hypothesis, Polignac's and Twin prime conjectures using Elementary-Emergent Fundamental Laws: Introduction incorporating Medical Materials BOOK SERIES 1*. Category: *Non-fiction*. Genre: *Popular Science*. Employing analogous terms such as 'Beautiful Mathematics' (for Completely Predictable problems) versus 'Sexy Mathematics' (for Incompletely Predictable problems) in this book should fulfil the designated role of "rewriting science" using friendly mathematical language for the general public. It essentially provide the "Introduction incorporating Medical Materials" to my successful proofs for Incompletely Predictable problems of Riemann hypothesis, Polignac's and Twin prime conjectures in 2019. Millennium Prize problem of Poincare conjecture was solved by Russian mathematician Grigori Perelman in 2003 but he famously decline the US$1 million prize awarded to him. Riemann hypothesis is another Millennium Prize problem. Perhaps a deserving comparison title for this book would be 'The Grand Design' (2010) written by renowned physicist and cosmologist Stephen Hawking (January 8, 1942 – March 14, 2018) and Leonard Mlodinow but that is for the public to determine. It is often said that Laws of Nature can be written using the language of mathematics. I always realize that doing mathematics not only relax my mind in time of stressful life events but also provide me with a second career pathway option. From early part of 2019, my medical career and earning sufficient income to financially support my family were in tatters – but that is another story in itself outlined later on.

1 The Big Picture of Fundamental Laws

"There are inherent Black and White Laws with complete accuracy applicable to Nonliving Things but only Black and White–like Laws with incomplete accuracy applicable to Living Things." By Dr. John Ting (June 1, 2019).

The above is my quote using Black and White (B&W) concept when applied to "Elementary" Nonliving Things and "Emergent" Living Things. We note mathematical-based proofs for Nonliving Things with simple or complex Elementary problems must be B&W correct (with absolute 100% certainty) but probability-based proofs for Living Things with simple or complex Emergent problems can only be B&W–like correct (with arbitrarily chosen level of statistical significance of less than and never equal to 100% certainty).

My two main research papers in 2019 on Riemann hypothesis, Polignac's and Twin prime conjectures were submitted to a mathematical journal from American Mathematical Society. These can be accessed as "original version" viXra papers listed as (Ting J. , Solving Incompletely Predictable problem Riemann hypothesis with Dirichlet Sigma-Power Law http://vixra.org/pdf/1903.0483v6.pdf, April 12, 2019) and (Ting J. , Solving Incompletely Predictable problems Polignac's and Twin Prime conjectures using Information-Complexity conservation http://vixra.org/pdf/1904.0214v3.pdf, April 26, 2019) [reproduced respectively in Appendix 1 and 2].

Errata Notice: There are mathematical errors present in (Ting J. , Rigorous Proof for Riemann Hypothesis Using the Novel Sigma-power Laws and Concepts from the Hybrid Method of Integer Sequence Classification https://doi.org/10.5539/jmr.v8n3p9, 2016) and (Ting J. , Key Role of Dimensional Analysis Homogeneity in Proving Riemann Hypothesis and Providing Explanations on the Closely Related Gram Points https://doi.org/10.5539/jmr.v8n4p1, 2016) as relevant equations which were [incorrectly] treated as 'equations' instead of being [correctly] treated as 'inequations'. These papers are corrected, refined and amalgamated together as (Ting J. , Solving Incompletely Predictable problem Riemann hypothesis with Dirichlet Sigma-Power Law http://vixra.org/pdf/1903.0483v6.pdf, April 12, 2019) with all equations, inequations and mathematical arguments peer reviewed to be correct and complete.

We compare for similarities and contrast for differences between Riemann hypothesis (RH) and Polignac's and Twin prime conjectures (P&TPC) below.

Firstly, RH and P&TPC are classified as Incompletely Predictable problems.

Secondly, to solve Completely Predictable problems require the simplicity of deriving their rigorous proofs based on simple properties whereas to solve Incompletely Predictable problems require the complexity of deriving their rigorous proofs based on complex properties ("meta-properties"). For Incompletely Predictable problems, the [few] complex properties are derived from the [many] underlying simple properties. As opposed to Completely Predictable problems such as dealing with even number gaps (= 2) or odd number gaps (= 2) endowed with simple properties which are easy to solve, dealing with RH and P&TPC which are Incompletely Predictable problems endowed with complex properties are mind-boggling hard to solve. RH and P&TPC are respectively connected with Incompletely Predictable entities nontrivial zeros (directly) derived from Riemann zeta function and prime numbers (directly) derived from Sieve of Eratosthenes. The act of explaining closely related Incompletely Predictable entities of the two types of Gram points which are (directly) derived [dependently] from Riemann zeta function should logically be classified as Incompletely Predictable problems. The Incompletely Predictable entities of composite numbers which are (indirectly) derived [dependently] from Sieve of Eratosthenes should logically be incorporated together with prime numbers to help solve P&TPC.

Thirdly, the two sets of prime and composite numbers exist at 'Numerical relationship

5

interface' with (solitary) "outlier" even prime number '2'; and the three sets of nontrivial zeros (or Gram[x=0,y=0] points), Gram[y=0] points (or 'traditional' Gram points) and Gram[x=0] points exist at 'Axes intercept relationship interface' with (solitary) "outlier" negative Gram[y=0] point.

Fourthly, deep seated connections exist between Riemann zeta function, $\zeta(s)$, and complete Set all prime numbers 2, 3, 5, 7, 11, 13,... [but not complete Subsets of prime numbers with each uniquely derived from prime gaps 1, 2, 4, 6, 8, 10,...]. The equivalent Euler product formula from Equation 1 in (Ting J. , Solving Incompletely Predictable problem Riemann hypothesis with Dirichlet Sigma-Power Law http://vixra.org/pdf/1903.0483v6.pdf, April 12, 2019) with product over all prime numbers [instead of summation over natural numbers] can also be used to represent $\zeta(s)$. Thus Set all prime numbers is intrinsically "inscribed" in $\zeta(s)$. Prime number theorem, fully delineated by prime counting function [denoted by $\pi(x)$], describes asymptotic distribution of all prime numbers among positive integers by formalizing intuitive idea that prime numbers become less common as they become larger through precisely quantifying rate at which this occurs using probability. Note: we must instead use Dirichlet eta function, $\eta(s)$, [which is the *proxy* function for $\zeta(s)$] to solve RH. Solving RH is instrumental in proving efficacy of techniques that estimate $\pi(x)$ efficiently thus confirming "best possible" bound for error ("smallest possible" error) of this theorem.

Fifthly, solving RH involves rigorously proving the complete Set nontrivial zeros [of known infinite magnitude] to be located on so-called critical line whereas solving P&TPC involves rigorously proving the existence of complete Set all prime numbers [of known infinite magnitude] to be constituted by Subsets of prime numbers [each proposed to be of infinite magnitude] uniquely derived from all even number prime gaps [proposed to be of infinite magnitude]. RH deals with 'Set' whereas P&TPC deals with 'Subsets'. Note that Polignac's conjecture concerns Subsets of prime numbers derived from all even number prime gaps 2, 4, 6, 8, 10,... but Twin prime conjecture concerns Subset of prime numbers derived only from even number prime gap 2. Thus

6

the later conjecture is intrinsically just part of the former conjecture. From Set theory, the Sets and Subsets of prime numbers will always comply with 'well-ordering Principle' (which states that every non-empty set of positive integers contains a least element) and 'pigeonhole principle' (which states that if n items are put into m containers with n>m, then at least one container must contain more than one item).

We now provide a Hierarchical Classification for Elementary-Emergent Fundamental Laws (EEFL). Implied by the definition for 'Fundamental Laws', then EEFL must by default be perfectly applicable to Terrestrial human beings on planet Earth (endowed with advanced civilization) and also Extraterrestrial alien beings on some hypothetical remote planet (endowed with super-advanced civilization). Thus one could also appropriately coin our Fundamental Laws as Extraterrestrial-Terrestrial EEFL. In order of increasing complexity, we have the following Laws:

Law I	Simple Elementary Fundamental Law for "simple" Nonliving Things with simple properties
Law II	Complex Elementary Fundamental Law for "complex" Nonliving Things with complex properties
Law III	Simple Emergent Fundamental Law for "simple" Living Things with simple properties
Law IV	Complex Emergent Fundamental Law for "complex" Living Things with complex properties

Solving Completely Predictable and Inompletely Predictable problems:

| Solving Completely Predictable problems in both Simple 'Nonliving' Elementary and 'Living' Emergent cases | *Many* Simple properties →[Simple Elementary and Emergent Solutions] |
| Solving Incompletely Predictable problems in both Complex 'Nonliving' Elementary and 'Living' Emergent cases | *Many* Simple properties →*Few* Complex properties →[Complex Elementary and Emergent Solutions] |

The three types of entities:

Type I Entities	Completely Unpredictable entities
Type II Entities	Completely Predictable entities
Type III Entities	Incompletely Predictable entities

Type I Entities occur purely as totally random physical processes in nature e.g. radioactive decay is a stochastic (random) process occurring at level of single atoms. According to Quantum theory, it is impossible to predict when a particular atom will decay regardless of how long the atom has existed. However for a collection of atoms, expected decay rate is characterized in terms of their measured decay constants or half-lives.

The location-based definitions for Type II Entities and Type III Entities are:

Completely Predictable (Type II) Entities	Locationally defined as entities whose position is **independently** determined by **simple** calculations using simple equation or algorithm **without** needing to know related positions of all preceding entities in neighborhood.
Incompletely Predictable (Type III) Entities	Locationally defined as entities whose position is **dependently** determined by **complex** calculations using complex equation or algorithm **with** needing to know related positions of all preceding entities in neighborhood.

Postulated association between Entities and Laws:

Law I is obeyed by Type II Entities	e.g. (simple Nonliving Thing) even and odd numbers with even number and odd number gaps.
Law II is obeyed by Type III Entities	e.g. (complex Nonliving Thing) nontrivial zeros related to RH, and prime numbers related to P&TPC.
Law III is obeyed by Type	e.g. (simple Living Thing) human heart as an organ manifesting hemodynamic and electrical properties.

II + Type III Entities	
Law IV is obeyed by Type I + Type II + Type III Entities	e.g. (complex Living Thing) human brain which is often dubbed "the most complex structure in the universe" manifesting a whole range of neuro-psychological and neuro-psychiatric properties.

Human heart can simplistically be thought of having a "plumbing system" consisting of heart muscle pump, coronary arteries and cardiac valves; and an "electrical system" consisting of specialized heart muscle cells giving rise to pacemakers and electrical conduction pathways & networks.

Human brain is the most complex organ in human body. It produces our every thought, action, memory, feeling and experience of the world. It consists of jelly-like mass of tissue weighing around 1.4 kilograms, and contains a staggering one hundred billion nerve cells (neurons).

The complexity of the connectivity between these cells is mind-blowing with each neuron making contact with thousands or even tens of thousands of others, via tiny structures called synapses. Our brains form about a million new connections per second. Our conscious mind commands and our subconscious mind obeys. Thus, our subconscious mind is an unquestioning servant that works day and night to make our behavior fits a pattern consistent with our emotionalized thoughts, hopes, and desires. The pattern and strength of the connections is constantly changing and no two brains are alike. It is in these changing connections that memories are stored, subconscious mind operate, habits learned and personalities shaped by reinforcing certain patterns of brain activity, and losing others.

Structurally, the human brain contains "grey matter" and "white matter". The grey matter is the cell bodies of the neurons, while the white matter is the branching network of thread-like tendrils called dendrites and axons that spread out from the cell bodies to connect to other neurons. However, the human brain also has another even more numerous type of cell called glial cells. These outnumber neurons about ten times over. Once thought to be support cells, they are now known to amplify neural signals and to be as important as neurons in mental calculations.

In summary, the human brain manifest Natural Intelligence, consciousness, self-awareness, memory; mental illness such as anxiety, depression, schizophrenia; "dark triad" of personality consisting of three negative traits [viz. the tendency to manipulate others (Machiavellianism), seek admiration and special treatment (narcissism), and to be callous and insensitive (psychopathy)]; and "light triad" of personality consisting of three positive traits [viz. the opposite of Machiavellianism (Kantianism), valuing dignity and worth of each individual person (humanism), and believing that people are fundamentally good (Faith in humanity)].

Artificial Intelligence (AI) in Nonliving Things can be regarded as human endeavor to simulate Natural Intelligence in Living Things using powerful computers such as super-computers or quantum computers. DNA is a double helix, while RNA is a single helix. Both have sets of nucleotides that contain genetic information. DNA is a molecule that contains instructions for Living Things to be born, mature, reproduce, and died.

One would commonly concur that there are 'Simple' Living Things such as bacteria without brain and 'Complex' Living Things such as intelligent human with highly developed brain. The dividing line between Living Things and Nonliving Things is that the former is "powered" by DNA with an important implication that Natural Intelligence, consciousness and self-awareness can only be "powered" by DNA [which are organic]. Then by reasonable assumption, properties such as consciousness and self-awareness can never be present in AI created using computers [which are inorganic].

Creationism versus Evolution debate for **Nonliving Things** (obeying Law I and Law II) giving rise to **Living Things** (obeying Law III and Law IV):

Process of Creationism	Process of Evolution
Occur over approximately two thousand years or so. Associated with major religions e.g. Islam and Christianity.	Occur over millions or billions of years. Associated with atheism.

People from most western countries generally embrace Christianity. To avoid conflicts, humanity must respect the freedom to practice [or not practice] all different religions. How can we reconcile the huge time discrepancy noted above between the process of creationism and evolution? A controversial thought is perhaps the process of "simplified" evolution with natural selection (survival of the fittest) and adaptation as plausible mechanisms occurs in both Simple and Complex Living Things on short, medium and long term scale in past, present and future. Complex Living Things with brains can only arise through creationism. In particular, the complex neuronal brain tissue can only be "created" by God and cannot "evolve" from simple living tissue over the four eons of geologic time scale.

The following are subjective comments. For Nonliving Things, we would intuitively associate performing calculation $2 + 3 = 3 + 2 = 5$ as a simple case of elementary Completely Predictable problem; and solving RH and P&TPC as complex cases of elementary Incompletely Predictable problems. For Living Things, we intuitively associate analyzing human heart as a simple case of emergent Completely Predictable problem; and analyzing human brain as a complex case of emergent Incompletely Predictable problem. Because mathematical language for describing complex Incompletely Predictable problems in Nonliving Things (such as weather forecasting) and Living Things (such as determining neurophysiology of human memory) are convoluted, we can only ever obtain approximate models of these problems.

Riemann hypothesis was proposed by famous German mathematician Bernhard Riemann (17 September 17, 1826 – July 20, 1866) in 1859. Twin prime and Polignac's conjectures were respectively proposed by French mathematician Alphonse de Polignac (1826 – 1863) in 1846 and 1849. In this chapter, I offer my personal opinion with some bold statements on why intractable open & Incompletely Predictable problems in Number theory of Riemann hypothesis (RH), Polignac's and Twin prime conjectures (P&TPC) have previously not been solved for over 150 years. The previous chapter outlining complicated similarities and differences between these open problems will already provide good insight why delay in solving them occur.

Perhaps most mathematicians have a weird sense of humor. My task to explain this delay is simplified using quirky terms 'Beautiful Mathematics' (BM) and 'Sexy Mathematics' (SM) to provide succinct mental pictures thus promoting optimal understanding by the general public:

Beautiful Mathematics (BM)	Mental picture for Completely Predictable problems which are **easy** to solve requiring BM which involves analyzing their (intrinsic) **simple** properties.
Sexy Mathematics (SM)	Mental picture for Incompletely Predictable problems which are **difficult** to solve requiring SM which involves analyzing their (intrinsic) **complex** properties

With the [few] complex properties derived from the underlying [many] simple properties in Incompletely Predictable problems, the caveat is that only through correctly analyzing these complex properties (or "meta-properties") will we ever obtain their complete solutions. Underlying simple properties in Incompletely Predictable problems can be falsely perceived to be complex properties. Complex properties in Incompletely Predictable problems can be hidden away in some subtle manner. Thus actual complex properties are notoriously difficult to correctly decipher with mathematicians frequently **barking up the wrong tree** in Incompletely Predictable problems. Another mistake is obtaining or claiming **'pseudo-proof'** whereby mathematicians would [incorrectly] utilize "manifestations" of relevant complex equations or complex algorithms from the Incompletely Predictable problems and compare these "manifestations" to some other *seemingly-related* Incompletely Predictable problems that were successfully solved in the past.

In other words for Incompletely Predictable problems, **barking up the wrong tree** is the equivalent of mathematician [incorrectly] analyzing the "beautiful tree" with simple properties using BM while **barking up the right tree** is the equivalent of mathematician [correctly] analyzing the "sexy tree" with complex properties using SM. Figuratively speaking, there are many "beautiful trees" choices but only a few "sexy trees" choices. So what actually is this so-called SM for RH and P&TPC? The answer is illustrated using the "mathematical impasse" phenomenon for Completely Predictable problem involving even numbers and their gaps (in Diagram 1) and for Incompletely Predictable problem involving prime numbers and their gaps (in Diagram 2).

Diagram 1. Legend: Even numbers = E, even number gaps = eGap.

E	2		4		6		8		10		12	
eGap		2		2		2		2		2		2	

Question: Prove the proposal that even number gaps are always constant and non-varying. Answer: Finite calculations shown in Diagram 1 depict and support even number gaps [= 2] is constant and non-varying but even numbers are infinite in magnitude requiring an infinite number of calculations ("mathematical impasse") in order to show these gaps will always be constant and non-varying. Obtaining rigorous proof then consist of recognizing this as Completely Predictable problem deriving a **Completely Predictable 'non-varying' equation** for calculating all even numbers which will [intrinsically] contain simple property "all even number gaps = 2". This equation is literally the 'Simple Container' containing all even numbers.

Diagram 2. Legend: Prime numbers = P, prime number gaps = pGap.

P	2		3		5		7		11		13		
pGap		1		2		2		4		2		4	

Question: Prove the proposal that apart from first prime number gap [=1] followed by next two consecutive prime number gaps [= 2], prime number gaps are always even numbers and varying. Answer: Finite calculations shown in Diagram 2 depict and support prime number gaps [after the third one] are even numbers and varying but prime numbers are infinite in magnitude requiring an infinite number of calculations ("mathematical impasse") in order to show these gaps will always be even numbers and varying. Obtaining rigorous proof then consist of recognizing this as Incompletely Predictable problem deriving an **Incompletely Predictable 'varying' equation** for calculating all prime numbers which will [intrinsically] contain complex property "all prime gaps are even numbers and perpetually varying". This equation is literally the 'Complex Container' containing all prime numbers.

As outlined in (Ting J. , Solving Incompletely Predictable problem Riemann hypothesis with Dirichlet Sigma-Power Law http://vixra.org/pdf/1903.0483v6.pdf, April 12, 2019) containing rigorous proof for RH and explaining Gram points, our 'overall' complex properties consist of three variants of Dirichlet Sigma-Power Laws precisely manifesting the required exact and inexact Dimensional analysis homogeneity. These novel Laws are derived from Dirichlet eta function, the proxy function for Riemann zeta function.

As outlined in (Ting J. , Solving Incompletely Predictable problems Polignac's and Twin Prime conjectures using Information-Complexity conservation http://vixra.org/pdf/1904.0214v3.pdf, April 26, 2019) containing rigorous proofs for P&TPC, our 'overall' complex properties consist of Plus-Minus Gap 2 Composite Number Alternating Law being precisely obeyed by all even number prime gaps apart from first even number prime gap precisely obeying Plus Gap 2 Composite Number Continuous Law. These Laws are derived using the novel research method Information-Complexity conservation.

Finally, another useful mental picture on why Incompletely Predictable problems such as Riemann hypothesis, Polignac's and Twin prime conjectures are so difficult to solve is that they require complex mathematical arguments belonging to '**Special-Class-of-Mathematical-Problems with Solitary-Proof-Solution**' whereas Completely Predictable problems such as proving even number gaps = 2 and odd number gaps = 2 only require simple mathematical arguments based on mathematical calculus or geometrical gradient method belonging to '**General-Class-of-Mathematical-Problems with Multiple-Proof-Solutions**'. Closely related Completely Predictable problems can easily be "separately" and "independently" solved whereas closely related Incompletely Predictable problems have to be "combined together" and "dependently" solved with difficulty.

I have practiced in the specialty field of Anesthesia, Intensive Care, Pain Medicine, Medicinal cannabis, and Addiction Medicine. When I was training in Anesthesia from 2009 to 2013, I was often told that the ability to effectively communicate to my patients is paramount for good patient care. Useful idiom: Patients 1st, Doctors and Nurses 2nd, Administration and Regulatory Body 3rd. To many, mathematical literature is often seen as an impenetrable wall of logic, symbols and formulas. I recommend this book to readers who want to get a meaningful glimpse of what is behind the wall and how the wall can be penetrated. I endeavor to write this book describing the artistic, creative and human, and spiritual aspect of mathematical enterprise.

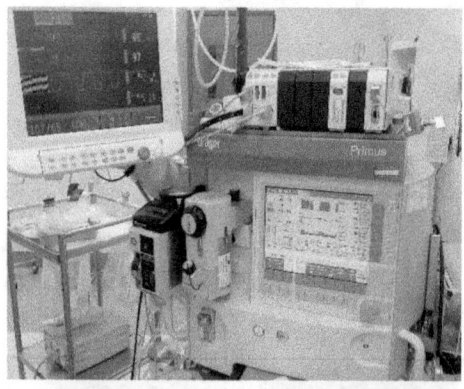

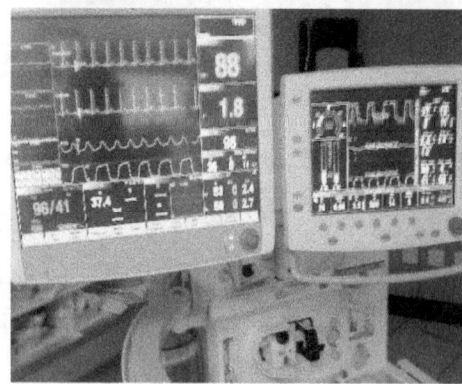

With three older sisters and two younger brothers, I am the eldest son of a domineering father. My mother died from a stroke at age 75 in 2016. Being raised on a rigmarole of high expectations and little praise, I developed an unwanted psychological trait with negative consequences but I resolved this issue in Year 2000 by initiating good father-son relationship.

On Monday May 14, 2012 my youngest daughter Jelena (meaning 'Shinning Light' in Russian) was born 13 weeks premature with a tiny birth weight of 1010 grams (2. 2 pounds). She spent 7 weeks in Neonatal Intensive Care Unit.

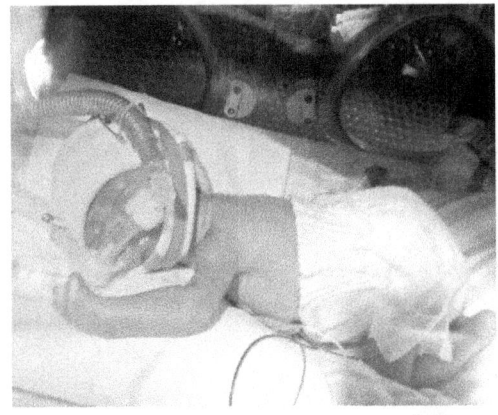 Remifentanil is a potent, short-acting synthetic opioid analgesic drug with an effective biological half-life of 3 to 10 minutes. As this useful drug is esterase metabolized, it is not dependent on the immature liver enzymes for metabolism. Therefore its theoretical advantage is that it provide the superior analgesia of an opioid without causing prolonged respiratory depression.

There is a popular saying 'Medicine is an inexact science'. Neonates do feel pain and require analgesic relief. For a good cause, my wife Jocelyn and I enrolled Jelena in the **premi-remi study** on May 21, 2012 at 28.2 weeks gestation during her PICC line insertion procedure in left ankle saphenous vein. With primary aim to determine efficacy of remifentanil infusion for alleviating pain in neonates requiring insertion of central venous lines for their medical care, this study is similar to the study "Remifentanil for percutaneous intravenous central catheter placement in preterm infant: a randomized controlled trial" by (Lago P, 2008). Thus Jelena became part of history contributing data as one of the recruited neonatal subjects in this study based on randomized double-blind controlled clinical trial.

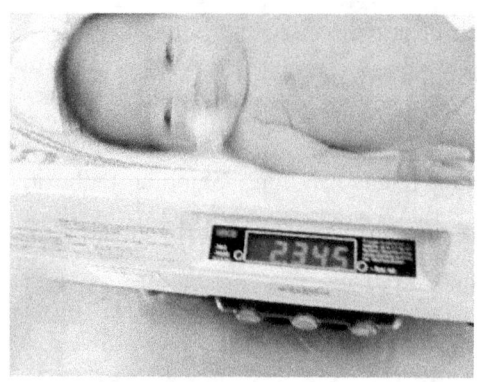

With a feeding tube up her nose, Jelena weight was exactly 2345 grams on July 13, 2012.

Jelena by her first birthday is an inspiration for all and a fighter developing into a normal healthy child. It is a no-brainer I will always be proud of her.

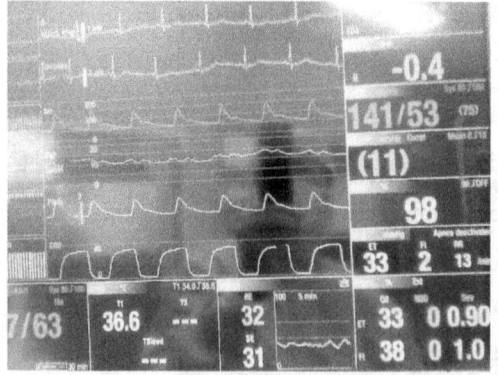

I have to stop my full time Anesthesia career in 2013 to help my wife look after my five children. I continue to do some

Anesthesia in rural parts of Australia until 2016. At beginning of 2019, my medical career and livelihood were in jeopardy. I can hardly support my family financially as I am unable to work casually in general practice due to regulatory requirements. These mainly arise from unnecessary allegations levelled against me out of practicing Addiction Medicine to combat drug dependency illness in a chaotic *en masse* manner for up to 123 difficult patients in 2017 and 2018 when their elderly doctor who has been practicing Addiction Medicine for large scale up to 150 patients and for more than 10 years was *forcibly* retired in October 2017.

This elderly doctor was [incorrectly] using large doses of multiple Benzodiazepine drugs for [misguided] 'opioid-sparing' role resulting in a number of drug overdose cases with harmful & fatal consequences. Through negotiation, I was asked by the relevant health department to initially "cover" for this doctor and subsequently "take over" care of these patients.

With my Opioid Replacement license obtained on June 10, 2017 and my extensive pharmacology knowledge gained from qualifications and experiences in Anesthesia, Intensive Care and Pain Medicine from 2009 to 2013; I commence in July 2017 to see this large group of patients under "quota of 100 patients" maiden approval for July 1, 2017 to June 30, 2018 period [instead of usually starting off with "quota of 20 patients" approval]. This large "quota of 100 patients" was furthermore [illegitimately] back-dated on October 10, 2017 to that effect by this health department.

On February 8, 2019 this health department [falsify] sending me a crucial email containing an important attachment; and furthermore on a separate February 13, 2019 email made a [false] admission that the alleged February 8, 2019 email was actually sent to me. This act was initially carried out on February 8, 2019 via this department intentional and unlawful [impersonating] use of incorrect jcting@hotmail.com instead of my correct jycting@hotmail.com email address. In addition, the relevant senior position person involved from this department made a number of false allegations against me about my confidential patient care not adhering to required regulations. All above *unfair plays* were elegantly proven by me. These breach the desperately needed shared responsibilities and mutual trust between regulatory body and doctor when working as a team to effect "harm minimization" or optimize "harm : benefit ratio" for patient care.

I have always provide safe treatments for these patients but I fail to complete some mandatory paper-works due to factors such as time constraint and practice location change on December 4, 2017 resulting in nil access to previous medical records for these patients. I am grateful to tackle these trials and tribulations with faith in God and support from close friends & family.

Rebelling against the pressure-cooker life being a doctor, I have always been privileged to be involved in my spare-time hobby of solving open problems in Number theory of Riemann hypothesis, Polignac's and Twin prime conjectures. I estimate it took me about three years from March 2016 to April 2019 to obtain rigorous proofs for these three open problems.

Finally, here is my choice for top three Mega-achievements by mankind:

(1) American astronaut and aeronautical engineer Neil Armstrong (August 5, 1930 – August 25, 2012) was the first person to walk on the Moon on July 21, 1969 and spoke the now-famous words, "That's one small step for [a] man, one giant leap for mankind."

(2) NASA Voyager 1 and Voyager 2 space probes both launched in 1977 to study the outer planets with computing power available in these probes less than that in a modern mobile phone.

(3) People's Republic of China as economic superpower with market economy

currently calculated in 2019 as the world's second largest economy by nominal gross domestic product (GDP) and the world's largest economy by purchasing power parity. Until 2015, China was the world's fastest-growing major economy with growth rates averaging 6% over 30 years. She is the world's largest manufacturing economy and trading nation; and also the world's fastest-growing consumer market and second-largest importer of goods. She is a net importer of services products.

4 Incompletely Predictable numbers, Gram points & A228186 (plus Medical narcotics & benzodiazepines)

Completely Unpredictable numbers arising from totally chaotic physical processes in nature give rise to countable infinite set (CIS) of measured true random numbers. In this sense, computational pseudorandom number generator employing deterministic logic cannot be a source for true random numbers. There are two types of Predictable number: CIS of Completely, and CIS of Incompletely Predictable numbers with former "contained" in [Simple Container] equation or algorithm obeying Simple Elementary Fundamental Laws, and later "contained" in [Complex Container] complex equation or algorithm obeying Complex Elementary Fundamental Laws.

Prime numbers behave pseudorandomly with a strange mixture of order (structure) and chaos. It is with this pseudorandomly behavior that prime [and composite] numbers are regarded as 'Pseudorandom numbers'. We allude readers here that usage of the term 'Pseudorandom number' is deemed to be synonymous with usage of our devised term 'Incompletely Predictable number'. The concepts behind 'chaos' and 'order' in prime numbers as Pseudorandom or Incompletely Predictable numbers are (respectively) exemplified using the following popular quotes obtained from the May 5, 1975 inaugural lecture 'The First 50 Million Prime Numbers' at Bonn University by mathematician Don Zagier. These two quotes are "...despite their simple definition and role as the building blocks of the natural numbers, the prime numbers belong to the most arbitrary and ornery objects studied by mathematicians: they grow like weeds among the natural numbers, seeming to obey no other law than that of chance, and nobody can predict where the next one will sprout" and "...the prime numbers exhibit stunning regularity, that there are laws governing their behavior, and that they obey these laws with almost military precision".

A Completely, and Incompletely Predictable number is location-defined as a number whose position is independently determined by simple calculations using simple equation/algorithm without, and dependently determined by complex calculations using complex equation/algorithm with needing to know related positions of all preceding numbers in its neighborhood. Both types of Predictable number exist as either rational [integers or fractions of two integers] numbers (**Q**) or irrational [algebraic or transcendental] numbers (**R − Q**). A well-defined set of **R − Q** will twice obey relevant location definition in CIS **R − Q** themselves and in CIS numerical digits after decimal point of each **R − Q**.

97 is an Incompletely Predictable number whose precise position is determined by computing positions of all preceding 24 prime numbers (**P**) using complex algorithm Sieve of Eratosthenes to conclude that 97 is the 25th **P**. Calculated using simple algorithm, 97 is also [i = (97+1)/2] 49th odd number (**O**) which is a Completely Predictable number. 98 & 99 are respectively [i = 98/2] 49th even number (**E**) & [i = (99+1)/2] 50th **O** which are Completely Predictable numbers calculated using simple algorithm. Determined indirectly using complex algorithm Sieve of Erastosthenes, 98 & 99 are respectively also the 72nd & 73rd composite numbers (**C**) which are Incompletely Predictable numbers.

Computing Riemann zeta function (or more specifically its *proxy* Dirichlet eta function) and Sieve of Eratosthenes will, respectively, supply Incompletely Predictable nontrivial zeros, Gram[y=0] & Gram[x=0] points and **P** & **C**. CIS of nontrivial zeros (denoted by imaginary part parameter t) = CIS of transcendental numbers = 14.134725, 21.022040, 25.010858, 30.424876, 32.935062, 37.586178,... [rounded off to six decimal places]. CIS of all **P** = Countable Finite Set (CFS) of all even **P** + CIS of all odd **P** = 2, 3, 5, 7, 11, 13,... whereby **P** '2' when treated as **E** is also regarded as a Completely Predictable number.

The three sets of nontrivial zeros, Gram[y=0] points and Gram[x=0] points, respectively, will dependently constitute three sets of Origin intercepts (or simultaneous x- & y-axes intercepts), x-axis intercepts and y-axis intercepts. Traditional 'Gram points' (or Gram[y=0] points) are relevant x-axis intercepts with choice of index 'n' for 'Gram points' historically chosen such that first 'Gram point' corresponds to t value which is larger than (first) nontrivial zero located at t = 14.134725. By convention, first six Gram[y=0] points will occur with following values [rounded off to six decimal places]: at n = -3, t = 0; at n = -2, t = 3.436218; at n = -1, t = 9.666908; at n = 0, t = 17.845599; at n = 1, t = 23.170282; at n = 2, t = 27.670182.

The two sets of **P** 2, 3, 5, 7, 11, 13,... and **C** 4, 6, 8, 9, 10, 12,... will dependently constitute set of natural numbers (**N**) 1, 2, 3, 4, 5, 6,... minus first **N** '1'. Whole numbers (**W**) = **N** plus '0'. '0' & '1' are special numbers being neither **P** nor **C** as they represent nothingness (zero) and wholeness (one), and the idea of having factors for '0' & '1' is meaningless. Treating '0' & '1' here as Completely or Incompletely Predictable numbers is also meaningless.

CIS of numbers derived from well-defined simple/complex algorithms or equations are "dual numbers" displayed as Completely & Incompletely Predictable number. Examples of **Q** '2' as **P** (& **E**), '97' as **P** (& **O**), '98' as **C** (& **E**) and '99' as **C** (& **O**) are described above. Examples of **R** -- **Q** are described next. First & only negative Gram[y=0] point (by convention at n = -3) with Completely Predictable y = 0 value is obtained by substituting Completely Predictable t = 0 resulting in $\zeta(\frac{1}{2} + it) = \zeta(\frac{1}{2}) = -1.4603545$, an Incompletely Predictable transcendental number [rounded off to seven decimal places] calculated as a limit similar to limit for Euler-Mascheroni constant or Euler gamma – its precise (1st) position can only be determined by computing positions of all preceding (nil) Gram[y=0] points. With exception of this first Gram[y=0] point, all t values from Gram[x=0] points, Gram[y=0] points, and nontrivial zeros (Gram[x=0,y=0] points) are Incompletely Predictable transcendental numbers – these are respectively associated with Completely Predictable x = 0, y = 0, and simultaneous x = 0 & y = 0 values. First 'Gram point' (by convention at n = 0 & is

24

associated with Completely Predictable x = 0 value from Incompletely Predictable t = 17.845599 substitution) is actually the 4th Gram[y=0] point whose precise (4th) position can only be determined by computing positions of all preceding (three) Gram[y=0] points. First nontrivial zero associated with simultaneous x = 0 & y = 0 value [equating to $\zeta(s) = 0$ whereby s = σ + it = $\frac{1}{2}$ + it] is Completely Predictable occurring with Incompletely Predictable t = 14.134725 value substitution – its precise (1st) position can only be determined by computing positions of all preceding (nil) nontrivial zeros.

Countable finite set (CFS) of Completely Predictable simple properties intrinsically present in Simple Container simple equations or algorithms help us solve Completely Predictable problems containing countable infinite set (CIS) of Completely Predictable numbers; whereas CFS of Completely Predictable complex properties intrinsically present in Complex Container complex equations or algorithms help us solve Incompletely Predictable problems containing CIS of Incompletely Predictable numbers.

Simple properties are inferred from a phrase like: "...the simple equation or algorithm by itself will intrinsically incorporate actual location [and actual positions] of all Completely Predictable numbers". Solving Completely Predictable problems endowed with simple properties which are amendable to simple treatments using usual mathematical tools such as Calculus will result in their Simple Elementary Fundamental Laws-based solutions. Complex properties are inferred from a phrase like: "...the complex equation or algorithm by itself will intrinsically incorporate actual location [but not actual positions] of all Incompletely Predictable numbers". Solving Incompletely Predictable problems endowed with complex properties which are amendable to complex treatments using unusual mathematical tools such as Information-Complexity conservation and exact & inexact Dimensional analysis homogeneity as well as usual mathematical tools such as Calculus will result in their Complex Elementary Fundamental Laws-based solutions.

Consider x for real number (**R**) values > 1. Let y be Set **R** such that (Simple Container simple equation) y = 2x or y = 2x - 1. This Simple Container will "contain" the complete uncountable infinite set (UIS) **R** [straight line of infinite length] commencing from Cartesian point (x=1, y=2) or (x=1, y=1). Computing y = 2x or y = 2x - 1 an infinite number of times – a "mathematical impasse" – will not *per se* result in its Simple Elementary Fundamental Laws-based solution for gradient or slope = 2. This gradient (simple property) is obtained by trigonometrically calculating tangent of y = 2x or y = 2x - 1 straight line which = 2 or mathematically analyzing y = 2x or y = 2x - 1 equation using Differential Calculus viz. $\dfrac{dy}{dx} = \dfrac{d(2x)}{dx}$ or $\dfrac{d(2x-1)}{dx}$ = 2. Note: applying Integral Calculus from Fundamental Theorem of Calculus to y = 2x or y = 2x - 1 for interval [1, +∞], viz. $\displaystyle\int_{1}^{\infty}(2x)dx$ or $\displaystyle\int_{1}^{\infty}(2x-1)dx$ = $[x^2 + C]_1^{\infty}$ or $[x^2 - x + C]_1^{\infty} = (\infty^2 + C) - (1^2 + C)$ or $(\infty^2 - \infty + C) - (1^2 - 1 + C)$ result in Simple Elementary Fundamental Laws-based solution for area (simple property) of ∞ size enclosed by mentioned straight line & x-axis.

By considering x > 1 integer number (**Z**) values for Simple Container simple algorithm y = 2x or y = 2x - 1, we obtain "contained" complete Set **E** or Set **O**. Computing **E** or **O** infinitely often – a "mathematical impasse" – will not *per se* result in its Simple Elementary Fundamental Laws-based solution for gap between any two consecutive **E** (**E** gap) or **O** (**O** gap) will both = 2. This gradient-equivalent **E** gaps or **O** gaps (simple property) is obtained by transforming these algorithms from their "discrete" formats into equivalent "continuous" formats [viz. "discrete" Δx = 1 → "continuous" Δx = 0] resulting in their gradients using either tangent method or Differential Calculus method as per previous paragraph. Then **E** or **O** gaps, both = 2, is numerically identical and mathematically equivalent to relevant gradients, both also = 2. The same method of transforming "discrete" formats into "continuous" formats required to solve Riemann hypothesis involves applying Riemann integral to "discrete-like" simplified Dirichlet eta function (in summation format) to obtain "continuous-like" Dirichlet Sigma-Power Law (in integral format).

Similar to Incompletely Predictable 'varying gaps' [equivalent to 'varying gradients'] between consecutive **P** (**P** gaps) & consecutive **C** (**C** gaps), we have Incompletely Predictable 'varying gaps' [equivalent to 'varying gradients'] between consecutive nontrivial zeros (nontrivial zero gaps), consecutive Gram[y=0] points (Gram[y=0] points gaps) & consecutive Gram[x=0] points (Gram[x=0] points gaps). These 'varying gaps' or 'varying gradients' (complex properties) are geometrically related to different shapes and sizes of spirals as depicted in the figure below on Riemann zeta function. Note from this figure that mathematically defined nontrivial zeros [as t values obtained when setting either $\zeta(s) = 0$ or $\eta(s) = 0$] is exactly equivalent to geometrically defined Gram[x=0,y=0] points [as all the 'Origin' intercepts].

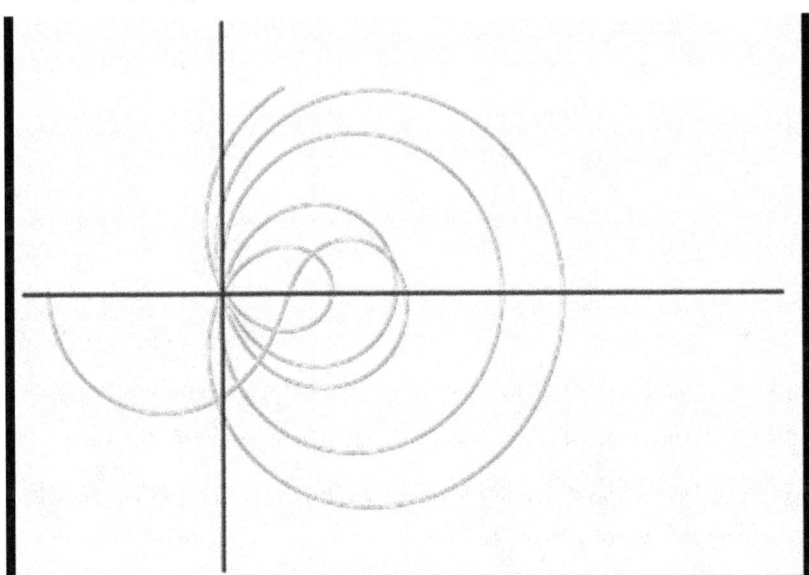

Gram's Law and traditional 'Gram points'

In mathematics, logarithm (log) is the inverse function to exponentiation. Log to base 10 is called the common logarithm and has many applications in science and engineering. Log to base e ($\approx$ 2.718) which is usually denoted by 'ln' [as it is in this book] is called the natural logarithm with widespread use in mathematics and physics because of its simpler derivative. Log to base 2 is called the binary logarithm and is commonly used in computer science.

Named after Danish mathematician Jørgen Pedersen Gram (June 27, 1850 – April 29, 1916), traditional 'Gram points' (or Gram[y=0] points which are **x-axis intercepts** shown in figure above) are other conjugate pairs values on critical line defined by $\text{Im}\{\zeta(\frac{1}{2} \pm it)\} = 0$. They obey Gram's Rule and Rosser's Rule with interesting characteristic properties as outlined by our brief exposition below. Z function is used to study Riemann zeta function on critical line. Defined in terms of Riemann-Siegel theta function & Riemann zeta function by $Z(t) = e^{i\theta(t)}\zeta(\frac{1}{2} + it)$ whereby $\theta(t) = \arg(\Gamma(\frac{(2it + 1)}{4})) - \frac{\ln\pi}{2}t$; it is also called Riemann-Siegel Z function, Riemann-Siegel zeta function, Hardy function, Hardy Z function, & Hardy zeta function. The algorithm to compute Z(t) is called Riemann-Siegel formula. Riemann zeta function on critical line, $\zeta(\frac{1}{2} + it)$, will be real when $\sin(\theta(t)) = 0$. Positive real values of t where this occurs are called 'Gram points' and can also be described as points where $\frac{\theta(t)}{\pi}$ is an integer. Real part of this function on critical line tends to be positive, while imaginary part alternates more regularly between positive & negative values. That means sign of Z(t) must be opposite to that of sine function most of the time, so one would expect nontrivial zeros of Z(t) to alternate with zeros of sine term, i.e. when θ takes on integer multiples of π. This turns out to hold most of the time and is known as Gram's Rule (Law) – a law which is violated infinitely often though. Thus Gram's Law is statement that nontrivial zeros of Z(t) alternate with 'Gram points'. 'Gram points' which satisfy Gram's Law are called 'good', while those that do not are called 'bad'. A Gram block is

28

an interval such that its very first & last points are good 'Gram points' and all 'Gram points' inside this interval are bad. Counting nontrivial zeros then reduces to counting all 'Gram points' where Gram's Law is satisfied and adding the count of nontrivial zeros inside each Gram block. With this process we do not have to locate nontrivial zeros, and we just have to accurately compute Z(t) to show that it changes sign.

Ratio Study and Inequations

A mathematical equation, containing one or more variables, is a statement that values of two ['left-hand side' (LHS) and 'right-hand side' (RHS)] mathematical expressions is related as equality: LHS = RHS; or as inequalities: LHS < RHS, LHS > RHS, LHS ≤ RHS, or LHS ≥ RHS. A ratio is one mathematical expression divided by another. The term 'unnecessary' Ratio (R) for any given equation is explained by two examples: (1) LHS=RHS and with rearrangement, 'unnecessary' R is given by $\frac{LHS}{RHS}=1$ or $\frac{RHS}{LHS}=1$; and (2) LHS>RHS and with rearrangement, 'unnecessary' R is given by $\frac{LHS}{RHS}>1$ or $\frac{RHS}{LHS}<1$.

Consider exponent y ∈ all **R** values & base x ∈ **R**≥0 values for mathematical expression $y^{\frac{x}{y}}$. Equations such as $x^1 = x$, $x^0 = 1$ & $0^y = 0$ are all valid. Simultaneously letting both x & y = 0 is an incorrect mathematical action because xy as function of two-variables is not continuous & is thus undefined at Origin. But if we elect to intentionally carry out this "balanced" action [equally] on x & y, we obtain (simple) inequation $0^0 \neq 1$ with associated perpetual obeyance of '=' equality symbol in x^y for all applicable **R** values except when both x & y = 0. The Number '1' value in this inequation is justified by two arguments: I. Limit of x^y value as both x & y tend to zero (from right) is 1 [thus fully satisfying criterion "x^y is right continuous at the Origin"]; and II. Expression x^y is product of x with itself y times [and thus x^0, the "empty product", should be 1 (no matter what value is given to x)].

29

Mathematical operator 'summation' must obey the law: We can break up a summation across a sum or difference but not across a product or quotient viz, factoring a sum of quotients into a corresponding quotient of sums is an incorrect mathematical action. But if we elect to carry out this action equally on LHS & RHS products or quotients in a suitable equation, we obtain two (unique) 'necessary' R denoted by R1 for LHS and R2 for RHS whereby R1 ≠ R2 relationship will always hold. We define 'Ratio Study' as intentionally performing this incorrect [but "balanced"] mathematical action on suitable equation [equivalent to one (non-unique) 'unnecessary' R] to obtain its inequation [equivalent to two (unique) 'necessary' R].

Let C denote complex numbers. Set C is a field (but not an ordered field). Thus it is not possible to define a relation between two given ($z1$ & $z2$) C as $z1 < z2$ since inequality operation here is not compatible with addition and multiplication. But performing Ratio Study to obtain inequations involving C does not involve defining a relation between two C.

A228186

The Hybrid method of Integer Sequence classification enables meaningful division of all integer sequences into either Hybrid or non-Hybrid integer sequences. My exotic A228186 (Ting J. , Hybrid integer sequence A228186 https://oeis.org/A228186. , August 15, 2013) integer sequence was published on The On-line Encyclopedia of Integer Sequences website in 2013. It is the first ever [infinite length] Hybrid integer sequence synthesized from Combinatorics Ratio. In 'Position i' notation, let $i = 0, 1, 2, 3, 4, 5,..., \infty$ be complete set of natural numbers.

A228186 "Greatest k > n such that ratio R < 2 is a maximum rational number with R

$$= \frac{\textit{Combinations With Repetition}}{\textit{Combinations Without Repetition}}"$$ is equal to [infinite length] non-Hybrid (usual garden-variety) integer sequence A100967 except for finite 21 'exceptional' terms

at Positions 0, 11, 13, 19, 21, 28, 30, 37, 39, 45, 50, 51, 52, 55, 57, 62, 66, 70, 73, 77, and 81 with their values given by relevant A100967 terms plus 1.

The first 49 terms [from Position 0 to Position 48] of A100967 (Noe, November 23, 2004) "Least k such that binomial(2k+1, k-n) ≥ binomial(2k, k)" are listed below: 3, 9, 18, 29, 44, 61, 81, 104, 130, 159, 191, 225, 263, 303, 347, 393, 442, 494, 549, 606, 667, 730, 797, 866, 938, 1013, 1091, 1172, 1255, 1342, 1431, 1524, 1619, 1717, 1818, 1922, 2029, 2138, 2251, 2366, 2485, 2606, 2730, 2857, 2987, 3119, 3255, 3394, and 3535. For those 21 'exceptional' terms: at Position 0, A228186 (= 4) is given by A100967 (= 3) + 1; at Position 11, A228186 (= 226) is given by A100967 (= 225) + 1; at Position 13, A228186 (= 304) is given by A100967 (= 303) + 1; at Position 19, A228186 (= 607) is given by A100967 (= 606) + 1; etc.

Here is a useful concept: Commencing from Position 0 onwards "in the limit" that this Position approaches 82, A228186 Hybrid integer sequence becomes (and is identical to) A100967 non-Hybrid integer sequence for all Positions ≥ 82.

Discovered in 2013, we will now devotedly explain A228186 in full details below due to its additional role in helping us understand the mathematical concept "in the limit...." employed in this book. A228186 has remarkable properties and whether there are other similar Hybrid Integer Sequences remains to be determined.

A228186: A unique Hybrid Integer Sequence

Hybrid method of Integer Sequence classification		
Main classes of integer sequence (infinite length)	Non-hybrid integer sequence (Class I)	Hybrid integer sequence (Class II)

	'Usual'	'Usual' and 'exceptional' terms (presumptively) derived from only 2 integer sequences in this classification system. 4 main varieties made up of either (i) [AXXXXXX (f(x)) AXXXXXX (f(y))] {near-identical integer sequences} with finite or infinite overlap, and (ii) [AXXXXXX (f(x)) AYYYYYY (f(y))] {non-identical integer sequences} with finite or infinite overlap.
Composition of relevant integer sequence	terms: made up of only 1 integer sequence: AXXXXX X (f(x)).	
Descriptive positions of 'exceptional' terms	Not applicable	'Exceptional' terms are either isolated, or interspersed with 'usual' terms. In principles all hybrid integer sequences with infinite overlap must be of the interspersed type, and those with finite overlap can be of either interspersed or isolated types.
Theoretical number of all possible subtypes	1	Ж$_6$ $\rightarrow$ 4

Legend: AXXXXXX and AYYYYYY represent individual integer sequence; f(x) and f(y) represent individual functional equation involving variable x and y respectively.

Ж In the real mathematical world, the theoretical number of all useful / practical (as opposed to all possible) subtypes of Class II Hybrid integer sequence should be limited to 4 (of the finite overlap variety) for the following reasoning. Notwithstanding the fact that the infinite overlap variety will result in the dilemma or arbitrariness of which near-identical or non-identical individual integer sequences should be designated as belonging to the 'usual' or 'exceptional' terms, their resultant hybrid integer sequence is simply a convenient mathematical combination / marriage of their individual integer sequence components with the (infinite) sum total of all its terms simply being exactly given by adding the (infinite) terms of each individual components. Reverse (and forward) modelling, akin to "engineering", of new hybrid integer sequences can take place at will by simply mathematically combining any two [or more] available infinite integer sequences – this do not result in any information gain from the resultant "totally predictable" hybrid integer sequences of this nature. On the contrary, the hybrid integer sequences of the finite overlap variety are "not totally predictable" with reverse engineering deemed to be extremely difficult if not impossible, and crucially it also simply CANNOT be fully described using only a solitary integer sequence, or using more than one in a simplistic "linear" manner. Also, the actual sum total of terms from this unique variety of integer sequence is less than the simple summation of the terms from the two involved infinite integer sequences if derived as such. In other words using the analogous language of thermodynamics, the increase in orderliness gleaned from the 'exceptional' terms being incorporated onto this hybrid integer sequence variety has resulted in the loss of information resulting from the remaining discarded 'exceptional' terms – this in essence equating to an increase in entropy.

Based on a new paradigm for our understanding of the classification of integer sequence, A228186 is a novel 'Hybrid integer sequence' with its specific denoted subtype. Our goal is to deem the associated mathematical work as ground-breaking research work in this relatively unknown scientific area in order to further the relevant knowledge here. The preliminary effort on our intended new, robust and functional classification method for integer sequence is outlined above. We eagerly await constructive criticisms from the wider scientific community.

Heavily illustrated with the help of Integer Sequence A000001 (Sloane, 1964), 'Number of groups of order n'. OEIS (founded in 1964 by N. J. A. Sloane); a much less satisfying alternative (but nevertheless insightful) classification of integer sequence is given in the following relevant paragraphs. The essential background preliminary arguments required to use the relevant keywords 'sequence' and 'pseudorandom' in "Pseudorandom infinite-length integer sequence" (PIIS) is predominantly based on the exclusion criteria and reproducibility principle – this exercise will justify the use of our coined PIIS and its role as an alternative classification of integer sequence. A sequence of integer numbers can theoretically be characterized by its finite-length [viz. finite-length integer sequence (FIS)] or infinite-length [viz. infinite-length integer sequence (IIS)], and their reproducibility [which loosely equates to predictability] or irreproducibility [which loosely equates to unpredictability].

Although useful as part of overall argument contributions, we will only briefly touch on the FIS type due to its relatively redundant nature. By logical deductions, a (literally) imaginary and non-existent FIS must always consist of a finite arbitrary set of chosen and memorized integers derived from a "perfect" random (or even a "slightly imperfect" pseudorandom) number generator, and that this set of integers can be mathematically proven [likely via various mathematical axioms] to be totally irreproducible, unpredictable & non-deterministic *per se* and also likewise by the same number generator – otherwise, this FIS will have all of its terms to be predictable and reproducible, and thus the total number of terms able to be extended to infinity (and belonging to the IIS class by default). In this setting, we can then in effect discard the "random infinite-length integer sequence" terminology not least due to its virtually useless mathematically descriptive label and revelation capacity. One can then argue that all "proper" integer sequences will always have to be IIS in character and thus also always having to be reproducible and infinite in nature.

Next, we can propose IIS to consist of two mutually exclusive classes viz. Class I: the traditional "non-pseudorandom infinite-length integer sequence" (which is the usual common-garden variety smooth integer sequences typified by many published integer sequences, and Class II: the contemporary "pseudorandom infinite-length integer sequence" – a simplistic & rarity example is currently typified by the solitary A228186 [this one] with A000001 from 'The Online Encyclopedia of Integer

34

Sequences' seemingly typifying a much more difficult example. The true nature of PIIS remains to be elucidated; for instance whether the pseudorandom components [as expounded below] of all existing PIIS can be of varying finite or infinite length and complexity in both their positions and values. Again using sheer logical deductions, finite pseudorandom components of PIIS commencing from its designated initial point will have to terminate at a subsequent designated point in the relevant sequence – otherwise if they fail to terminate, they will obviously have to be (by default) of the infinite pseudorandom components variety.

A typical definition of 'pseudorandomness' here is that, given one value of the sequence, finding the previous (or next) value is hard. Moreover, the task of finding all previous (or next) [infinite] values of the sequence would also take an infinite amount of time. However, our A228186 sequence is quite smooth and it should not be difficult to find terms in either direction [with the proviso] as long as n is not less than 82 – this also with the crucial contrasting implication that finding all the pseudorandom ('exceptional') terms/component contained within n = 0, 1,..., 81 would only take a finite amount of time. So we are also using the word 'pseudorandom' in a different manner in A228186 with the intended definition as follows. A228186 is pseudorandom in the sense that, given a sequence of N consecutive last bits of this sequence [n = 82, 83,..., ∞], it is extremely hard if not impossible to find an index starting a sequence with those final bits. An important corollary observation & caveat, based on the underlying arguments given above, is that when the pseudorandom components were to exist as infinite number of terms, they will still be reproducible and predictable [in the strictest sense of the words] despite hypothetically or conceivably needing an infinite amount of time to elucidate all of their infinite available pseudorandom terms. In addition, all pseudorandom terms in the integer sequences with infinite pseudorandom components will be viewed in its entirety as pseudorandom ('exceptional') terms always interspersed amongst the non-pseudorandom ('usual') terms. Thus by aesthetic reasoning alone, one would be tempted or inclined to suggest that pseudorandom components of the infinite type do not really exist in nature. However & perhaps paradoxically, nature is counter-intuitively and abundantly blessed with integer sequences having just such pseudorandom components of the infinite type – this being typified by A000001, formerly M0098 N0035, given as 0, 1, 1, 1, 2, 1, 2, 1, 5, 2, 2, 1, 5, 1, 2, 1, 14, 1, 5, 1, 5, 2, 2, 1, 15, 2, 2, 5, 4, 1, 4, 1, 51, 1, 2, 1, 14, 1, 2, 2, 14, 1, 6, 1, 4, 2, 2, 1, 52, 2, 5, 1, 5, 1, 15, 2, 13, 2, 2, 1, 13, 1, 2, 4, 267, 1, 4, 1, 5, 1, 4, 1, 50, 1, 2, 3, 4, 1, 6, 1, 52, 15, 2, 1, 15, 1, 2, 1, 12, 1, 10, 1, 4, 2,....

Features of A228186

A228186 is completely defined by the following mathematical statement: Greatest k > n such that ratio R < 2 is a maximum rational number with R =

$$\frac{Combinations\ with\ repetition}{Combinations\ without\ repetition}.$$ Least k > n such that binomial(2k+1, k-n-1) ≥ binomial(2k, k), or one more than this value if n is in [the 21 'exceptional' terms] {0, 11, 13, 19, 21, 28, 30, 37, 39, 45, 50, 51, 52, 55, 57, 62, 66, 70, 73, 77, 81} also completely define all terms in A228186.

At this point in time, to the best of our knowledge A228186 is the only known Class II Hybrid integer sequence of the specific subtype "[AXXXXXX(f(x)) AXXXXXX(f(y))] with finite overlap of the interspersed type". Each integer sequence made contributions to the 'usual' & 'exceptional' terms in the following manner & proportion: A100967(+0) occupying all consecutive positions when n is from 82 to ∞; and A100967(+1) & A100967(+0) respectfully occupying the designated positions for the 21 'exceptional' terms & 61 'usual' terms when n is from 0 to 81. [The convention employed here being A100967(+0) & A100967(+1) respectfully symbolizing add 0 and add 1 to every single A100967 term.] Based solely on these unique, finite and totally predictable pseudorandom 'exceptional' terms, we advocate A228186 as a novel true pseudorandom infinite-length integer sequence. The consequent role from this bold proposal on A228186 – correct to the best of our knowledge at the time of composing A228186 – and its highly significant connections with integer sequence A100967, combinatorics, and the mathematical fields of Chaos and Fractals could then be deemed to consist of ground-breaking research work in this relatively unknown scientific area. We will further elaborate all above points in materials below.

The topics of permutations (P) and combinations (C) come under combinatorics. P is an ordered C. There are two types of P and two types of C with their relevant formulae given below, given n = 0, 1, 2,..., ∞ [Symbol ! represent the factorial operation viz. the factorial of a positive integer n denoted by n! is the product of all positive integers less than or equal to n. Thus n! = n X (n-1) X (n-2) X (n-3) X ... X 3 X 2 X 1.]:

P with repetition: $k^{(n+2)}$ Equation (1)

P without repetition: $\dfrac{k!}{(k-n-2)!}$ Equation (2)

C with repetition: $\dfrac{(k+n+1)!}{(n+2)!(k-1)!}$ Equation (3)

C without repetition: $\dfrac{k!}{(n+2)!(k-n-2)!}$ Equation (4)

Numerically, (1) > (2) > (3) > (4) always holds true and R = $\dfrac{(3)}{(4)}$.

Least k > n such that binomial(2k+1, k-n-1) ≥ binomial(2k, k), or one more than this value if n is in {0, 11, 13, 19, 21, 28, 30, 37, 39, 45, 50, 51, 52, 55, 57, 62, 66, 70, 73, 77, 81} completely define all terms in A228186 [see Fig 1, Graphs related to A228186 below]. It consists of minority finite 21 'exceptional' terms (denoted by k+1) and majority infinite 'usual' terms (denoted by k). From the formula below for both 'usual' and 'exceptional' terms alike, we can also estimate n as approximately $0.83\sqrt{k}$ - 1 and $0.83\sqrt{(k+1)}$ - 1 if we know k and k+1 respectively.

"Greatest k > n such that R < 2 is a maximum [=
$$\frac{(k + n + 1)!(k - n - 2)!}{k!(k - 1)!}$$
< 2 is a maximum]" results in the complete set of A228186 and essentially represents a study on inherent properties (predominantly on mathematical patterns) for fractional part of the argument for R. These naturally present properties can intrinsically be determined graphically, by calculations, and by mathematical deductions based on the key concept "maxima R values < 2 equates to fract(R) being the maxima occurring at a particular k for 'usual' terms and k+1 for 'exceptional' terms" as further outlined below. This then will also provide sound explanations for the occurrence of actual number and position for all known 'exceptional' terms.

Binomial is Combinations without repetition. "Least k > n such that
$$\text{binomial(2k+1, k-n-1)} \geq \text{binomial(2k, k)} \left[= \frac{(2k + 1)!}{(k - n - 1)!(k + n + 2)!} \right.$$
$$\left. \geq \frac{(2k)!}{(k!)^2} \right]"$$
is the exact replica of A100967 integer sequence which was equivalently expressed previously as the inequality [in slightly different format here with using the notation: n = 0, 1, 2,..., ∞] "Greatest k > n such that binomial(2k+1, k-n-1) < binomial(2k, k)".

Intra-R analysis: Graphically plot fract(R) on y-axis versus k on x-axis for each n using integers from 0 onwards [see Fig 2, Graphs related to A228186]. For each graph with its finite number of peaks and troughs observed, maxima R values < 2 equates to fract(R) being the maxima occurring at a particular k for 'usual' terms, and k+1 for 'exceptional' terms respectively. As a real physical entity, this particular R value equates to the last (and highest) peak on Right Hand Side of x-axis; and its k or k+1 value is contained in A228186 for that particular n. The value k when [incorrectly] used for an 'exceptional' term will result in fract(R) being a minima with relatively much smaller value than this maxima value.

Inter-R analysis: Graphically plot each of the 2 individual data sets derived from fract(R using k) and fract(R using k+1) on y-axis versus n on x-axis for n from 0 onwards [see Fig 4, Graphs related to A228186]. For each of the distinct graphical line with its own peculiar finite number of peaks and troughs observed, fract(R using k) > fract(R using k+1) always holds true except at the 21 'exceptional' terms whereby fract(R using k) is a nadir value close to 0 and fract(R using k+1) is a zenith value close to 1. With progressively higher 'exceptional' terms, their fract(R using k+1) values will overall be monotonously rising steadily approaching a value of 1 (boundary condition), which then limit the total number of possible 'exceptional' terms. This intuitive mathematical deduction utilizing the imposed boundary condition supports our preposition that the number of 'exceptional' terms is finite.

In effect both the fract(R using k) and its closely-related fract(R using k+1), with a constant value difference of 1 amongst their corresponding terms, could be seen to reflect near-identical A228186 integer sequences. A228186 can also be perceived to consist of an infinite number of 'usual' (pure) terms tainted by a finite number of 'exceptional' (impure) terms, and can be eloquently designated by various nonlinear algorithms and formulae application thus manifesting the typical property of Chaos known as "deterministic

39

pseudorandomness". Resulting relevant graphs [see Fig 6, 8 & 11, Graphs related to A228186] employing various calculations based on A228186 will spontaneously manifest the typical property of Fractals known as "self-similarity" with simultaneous depiction of [opportunistic] recurrences in identical sets of the same 21 'exceptional' terms.

The 21 'exceptional' terms can be derived by employing the following steps. Perform an infinite number of iterations on formula $k = \text{Round}(0.3807 + 1.43869(n+1) + 1.44276(n+1)^2$ from $n = 0$ onwards while simultaneously substituting each n and its derived k value into R. It then follows that, respectively, the 'usual' [and 'exceptional'] terms will always be from maxima R values < 2 [and $\geq$ 2]; or equivalently expressed as its reciprocal, $\dfrac{1}{R}$, being from minima rational numbers with values ≥ 2 [and < 2]. The occurrence of any n and k with its calculated fract(R using k) being a minima close to 0 is the signature that this specific k is an 'exceptional' term with the corrected 'exceptional' term needing to be defined by k+1 instead. Alternatively, add 1 to each k value obtained from iterations on the formula and substitute using this (k+1) value instead. The graphical trend of the R obtained using this k+1 values will largely mirror but always be smaller than their corresponding counterparts using k values. The signature for an 'exceptional' term is then denoted by the calculated fract(R using k+1) having a maxima value close to 1 [but now with all other corrected 'usual' terms to be defined by k instead].

Denoted by the syntax product(f,var,a,b), this function returns the product of f where var is evaluated for all integers from a to b. Importantly, 'R using k' and 'R   using   k+1'   can   be   further   simplified   to

$$\frac{product(n,n,k+1,k+n+1)}{product(n,n,k-n-1,k-n,k-1)}$$ and

$$\frac{product(n,n,k+2,k+n+2)}{product(n,n,k-n,k)}$$ respectively, whereby k =

Round$(0.3807 + 1.43869(n+1) + 1.44276(n+1)^2$.

In summary, to the best of our knowledge A228186 represents a novel and freshly discovered Class II Hybrid integer sequence of the extremely specific subtype '[AXXXXXX(f(x)) AXXXXXX(f(y))] with finite overlap of the interspersed type' & a true 'Pseudorandom infinite-length integer sequence' based solely on its 21 unique pseudorandom 'exceptional' terms. We are strong advocates of this bold statement because of the following observations. Although the positions of the 21 'exceptional' terms are seemingly random in nature, they are also totally predictable. Moreover they only symbolize the tiny finite number of differences (of constant value 1) between two different combinatorics-flavored infinite series contained in A228186 and A100967 with their remaining infinite terms being exactly the same. Like A100967 (with one component from combinatorics, viz. Combinations without repetition), A228186 (with two components from combinatorics, viz. Combinations with and without repetition) is a genuine and real physical entity [albeit with differing number of combinatorics components] able to be vividly depicted graphically in their various formats. It constitutes a pioneer study on fractional part of the argument for ratio R, a rational number simply defined using two selected components from combinatorics, and should encourage future niche research studies into this previously little-known scientific area.

Finally, the pseudorandom nature of A228186 owes much of its explanations (and importantly, deep-seated connections) to certain non-

41

frivolous aspects derived from the ubiquitous and exciting mathematical fields of Chaos and Fractals – in particular and respectively, this refer to "deterministic pseudorandomness" and "self-similarity".

G r a p h s   r e l a t e d   t o   A 2 2 8 1 8 6

Figure 1 **Scatterplot of A228186(n)**

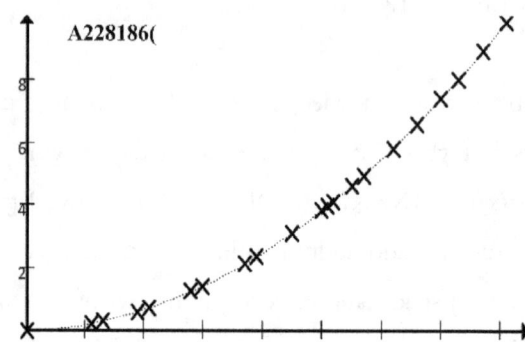

Legend: The 21 'Exceptional' terms denoted by X.

Figure 2

Intra R analysis: fract(R) vs k

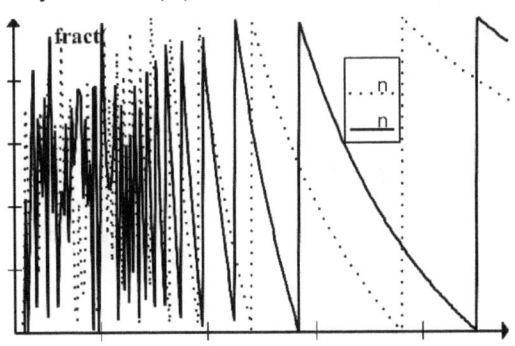

Figure 3

Inter-R analysis: R(k), R(k+1) vs n

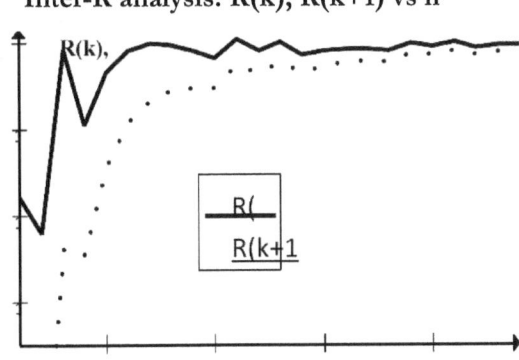

Figure 4

Inter-R analysis: fractR(k), fractR(k+1) vs n

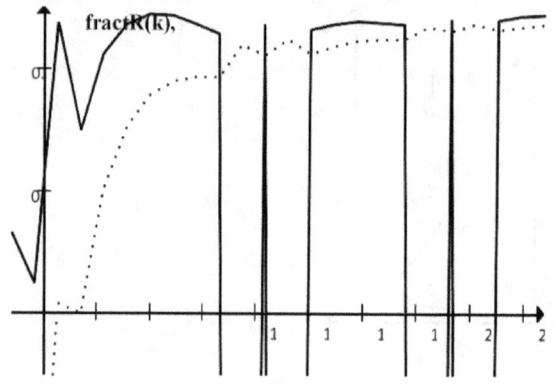

Legend: **fractR**(k)_____ **fractR**(k+1)......

Figure 5

Inter-R analysis: R(k)-R(k+1)

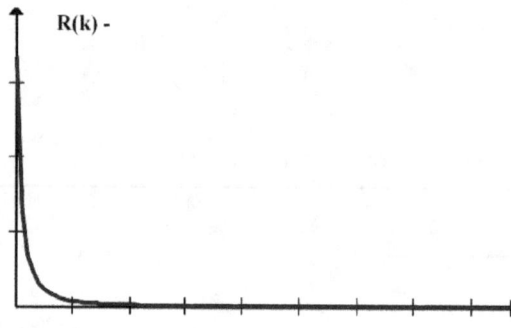

Figure 6

Inter-R analysis: fractR(k)-fractR(k+1)

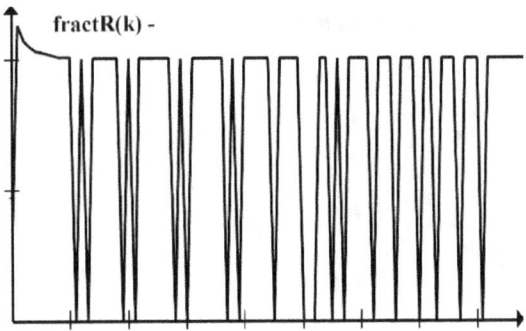

Figure 7

Inter-R analysis: DELTA R(k)

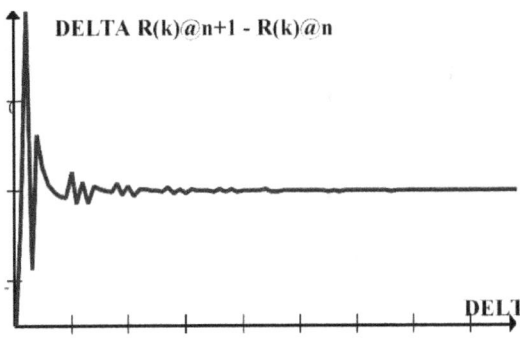

45

Figure 8

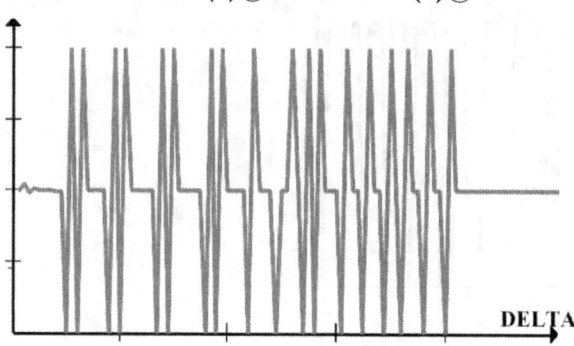

Inter-R analysis: DELTA fractR(k)

DELTA fractR(k)@n+1 - fractR(k)@n

DELTA

Figure 9

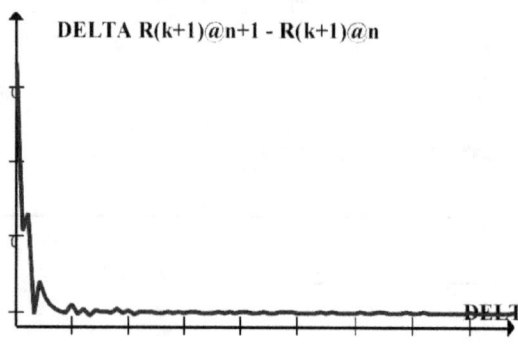

Inter-R analysis: DELTA R(k+1)

DELTA R(k+1)@n+1 - R(k+1)@n

DELT

Figure 10

Inter-R analysis: DELTA fractR(k+1)

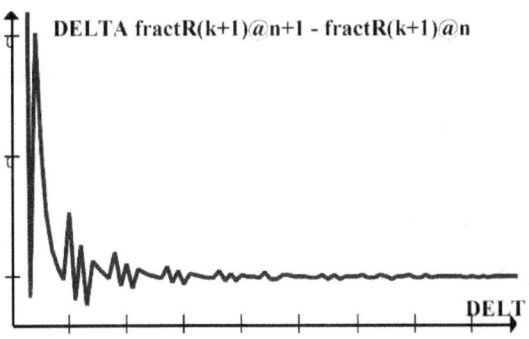

DELTA fractR(k+1)@n+1 - fractR(k+1)@n

DELT

Figure 11

Inter-R analysis: DELTA R(k) minus R(k+1) @n+1 & n

DELTA [fractR(k)@n+1 - R(k)@n] − [fractR(k+1)@n+1 - R(k)@n]

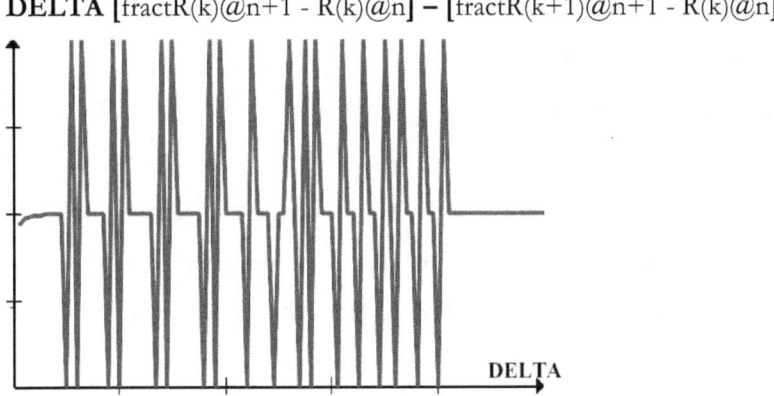

DELTA

To end this chapter, I provide the following interesting research ideas in relation to Medicine dubbed Paradoxical-to-Modern Science Law.

Paradoxical-to-Modern Science Law

A *paradox* is an argument that produces an inconsistency, typically within logic or common sense. Although most logical paradoxes are known to be invalid arguments, they can still be valuable in promoting critical thinking. Not surprisingly, some have revealed errors in definitions assumed to be rigorous, and have even caused axioms of mathematics and logic to be re-examined (e.g. Russell's paradox). Employing lateral thinking in an analogous manner, I devise the rule: In the event that initial seemingly *Paradoxical Practices and Phenomena* (PPP) can rigorously be validated by research with accompanying sound theoretical arguments, they are then able to be reclassified as final proven *Scientific Practices and Phenomena* (SPP).

We intuitively provide this rule as below:

Paradoxical-to-Modern Science Law

Paradoxical Science (*PPP*)

↓*Research validation* ↑(*Reverse direction*)

Modern Science (*SPP*)

Limbo Basket: Those Practices and Phenomena not able to be easily classified as either PPP or SPP on a temporary or permanent basis.

Legend: (***Reverse direction***) = current disproved or superseded SPP demoted to PPP

At my previous workplace in relevant medical fields, the diverse terms *Practices and Phenomena* could easily be referring to Anaesthesia and Intensive Care Practices and Phenomena. Thus, the relevant *Paradoxical Anaesthesia and Intensive care Practices and Phenomena* in question will become *Modern Anaesthesia and Intensive care Practices and Phenomena* when validated by research. Common sense will tell us here that especially with regards to *Practices*, many of these entities would be prone to be in the Limbo Basket of being neither classified as Paradoxical Practices nor Scientific Practices.

A somewhat related concept is 'Off-label drug use'. This is defined as use of pharmaceutical drugs for an unapproved indication or in an unapproved age group, dosage, or route of administration whereby the ability to prescribe drugs for uses beyond the officially approved indications is commonly carried out to good effect by healthcare providers. Indications for generic drug Aspirin are to reduce fever and relieve mild to moderate pain from conditions such as muscle aches, toothaches, common cold, and headaches, etc. It is also known to "thin the blood". An off-label drug use example would be the commonly accepted practice of giving 300 mg soluble Aspirin to a person suspected of having a heart attack whereby soluble Aspirin is not pharmaceutically registered for this purpose.

In contrast, 100 mg Aspirin commonly under brand-name 'Cartia' (endowed with enteric coating to lessen its Gastrointestinal side effect of causing peptic ulceration) is registered for this purpose under the approved indications for its use in known cardiovascular or cerebrovascular disease, as an antiplatelet agent for prophylaxis against acute myocardial infarction, unstable angina, transient ischaemic attack and cerebrovascular accident (stroke).

There are potential beneficial roles of occasional multiple narcotic use in modern pain management:

Point 1. Opioid analgesic drugs tend to exhibit incomplete cross-tolerance so that even when a patient has developed a high level of tolerance to one drug from the opioid class, they may find that a different opioid drug will still be effective – this is the concept/rationale for employing **Opioid Rotation.**

Point 2. The reasons for incomplete cross-tolerance is likely because of variations in opioid receptor affinity and occupancy levels at equianalgesic doses, as well as additional mechanisms of action possessed by some narcotic drugs such as NMDA antagonist action of Methadone, or SNRI activity of Tramadol or Tapentadol – the additional mechanisms of these drugs led to the potential use of **Multiple Opioids in acute or chronic pain scenario** on acute (short-term), subacute (medium-term), or chronic (long-term) basis. In comparison to Methadone (full agonist mu receptor with receptor affinity comparable to heroin), Buprenorphine has a ceiling effect to analgesia with "safer" less potential of respiratory depression due to partial mu-opioid receptor agonist (with receptor affinity >> Methadone/heroin but approximately = Nalaxone/Naltrexone) and fewer psychotomimetic & dysphoric effect due to kappa-opioid receptor antagonist. Note that Methadone and Buprenorphine are the two common Opioid Replacement drugs acceptably and legally used to control drug addiction in many societies.

Point 3. Drug dependent patients requiring controlled/restricted, for example, dangerous narcotic "S8 controlled drugs" (with or without significant acute or chronic pain component) should only be prescribed under a treating doctor with regulatory approval for that particular Drug X. Another nominated doctor in the same practice during the treating doctor's absence may prescribe Drug X. A doctor may judiciously prescribe Drug X or another "S4 restricted drug" (e.g. the Benzodiazepines) or S8 controlled drug" in an emergency setting for acute pain control whereby individual variation / pharmacokinetic / pharmacodynamic principles involved for the opioid-tolerant [as opposed to opioid-naïve] patient usually dictate the use of larger than normal doses of Drug X due to narcotic tolerance or larger than normal doses of another narcotic drug to competitively displace the original Drug X from mu-opioid receptors.

Point 4. Multi-modal approach to acute and chronic pain management may involve combining physical therapy (e.g. physiotherapy or Spinal Cord Stimulator for low back pain) and multiple drug therapy viz. time-limited narcotic trial / minimizing chronic narcotic use (as persistent non-cancer pain can have poor response to narcotic e.g. from (Farzana Mitra, January 15, 2013) *A feasibility study of transdermal buprenorphine versus transdermal fentanyl in the long-term management of persistent non-cancer pain.* Pain Medicine, 14 (1). pp. 75 – 83. https://doi.org/10.1111/pme.12011; only 11 - 13% have pain relief at 12 months from Fentanyl / Buprenorphine patches and 1/3 cease their use from side effect) and adding pain adjunct medications to narcotic such as tricyclic antidepressant Amitriptylline with Number Needed to Treat (NNT) 2.1, Pregabalin with NNT 4 – 6, Clonidine, Non-steroidal anti-inflammatory drugs (NSAIDS), Paracetamol, topical Lignocaine patch, and Capsaicin cream. NNT is the number of patients you need to treat to prevent one additional bad outcome (death, stroke, etc.). For example, if a drug has an NNT of 5, it means you have to treat 5 people with the drug to prevent one additional bad outcome.

Examples of short-term use are Fast-onset Alfentanil to blunt intubation response + later on add short acting Fentanyl or Morphine during General Anesthesia / Total Intravenous Anesthesia (TIVA) use of potent ultra-short acting Remifentanil + later on add Fentanyl or Morphine; Use non-narcotic drugs and/or a larger dose of non-Methadone or non-Suboxone narcotic drug(s) to compete with Methadone or Suboxone at the receptor level in emergency acute pain management. Use of Tramadol + Methadone is a chronic long term basis case example of managing patient's narcotic addiction + chronic pain from (for example) severe Rheumatoid Arthritis.

In the vast majority of cases, doctors should usually endorse / follow the general recommendation of using either solitary Suboxone or Methadone alone without adding any other narcotic in opioid dependent patients even when they have chronic pain. Adjusted for bio-availability, the Opioid Replacement equi-analgesic doses of Methadone & Buprenorphine as "Subutex" or "Suboxone" [Buprenorphine-Naloxone combination] prescribed for opioid dependent patients' narcotic addiction (Opioid-tolerant patients) are much higher than the doses of the same drugs used for acute (often Opioid-naïve patients) or chronic (often Opioid-tolerant patients) pain management. Example, calculations using the app 'Opioid Calculator': Opioid Replacement 18mg of Buprenorphine sublingually per day = oral Morphine Equivalent Daily Dose (oMEDD) of 720 mg per day. 40 mcg/hr (960 mcg or 0.9 mg/day) topical Buprenorphine (Norspan) patch = oMEDD of 80 mg/day only. Owing to the pharmacokinetic long half-life of Methadone, the time required to achieve dose stability varies from 35 to 325 h (13.5 days). Thus, oMEDD <30 mg/day led to oMEDD:Methadone ratio 2:1, oMEDD 31 to 99 led to oMEDD:Methadone ratio 4:1, oMEDD 100 to 299 led to oMEDD:Methadone ratio 8:1, oMEDD 300 to 499 led to oMEDD:Methadone ratio 12:1, oMEDD 500 to 999 led to oMEDD:Methadone ratio 15:1, oMEDD 1000 to 1200 led to oMEDD:Methadone ratio 20:1 – example, for subacute pain control Bradley Gardiner 80 mg Methadone per day = oMEDD 320 mg/day + 500mg per day Tapentadol = oMEDD 200mg/day + (say) 600 mg/day PRN Tramadol = oMEDD 120mg/day => Total oMEDD 640 mg/day [on weaning pathway for Tapentadol and Tramadol]. Other examples: Patients with chronic Low Back Pain and chronic neuropathic pancreatitis pain on opioid replacement Suboxone + "as required" PRN Tapentadol (Palexia) chronic use with approval from appropriate regulatory health organization.

Point 4. Opioid tolerant patients can get Opioid Withdrawal Syndrome such as increased pain, abdominal cramps, etc with insufficient narcotics and Opioid Intoxication of respiratory and CNS depression with excessive narcotics.

5 Axes Intercept Relationship Interface

In the Class of n-Variable Equations with n = 2 [which translate to 2-Variable Equations], when computationally depicted by 2-dimensional graphs with their x- and y-axes relevantly defined; they often have one or more points of intersection on (i) x-axis, and/or (ii) y-axis, and/or (iii) both x- and y-axes [formally known as the 'Origin']. The Origin, often labeled with capital letter 'O', is defined as the point where the vertical y-axis and the horizontal x-axis intersect each other. Not all functions, though, will have intercepts; which are where the graph crosses either the x-axis (viz. the x-axis intercept, often referred to as "zeros or roots of the equation"), or the y-axis (viz. the y-axis intercept), or both the x- and y-axes (viz. the Origin intercept).

There are eight possible Categories of Intercepts for 2-Variable Equations:

Category I Intercept: comprising of nil intercept

Category II Intercept: comprising of single x-axis intercept(s) only

Category III Intercept: comprising of single y-axis intercept(s) only

Category IV Intercept: comprising of single Origin intercept(s) only

Category V Intercept: comprising of double x- & y-axes intercept(s)

Category VI Intercept: comprising of double x-axis & Origin intercept(s)

Category VII Intercept: comprising of double y-axis & Origin intercept(s)

Category VIII Intercept: comprising of triple x-, y-axes & Origin intercept(s)

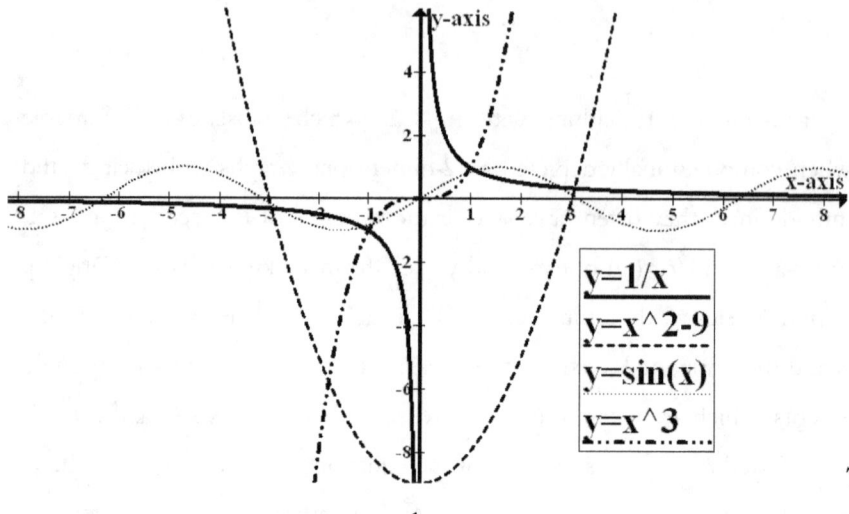

The above figure depicts simple formulae $y = \dfrac{1}{x}$, $y = x^2 - 9$, $y = \sin(x)$ and $y = x^3$ with their axes intercepts.

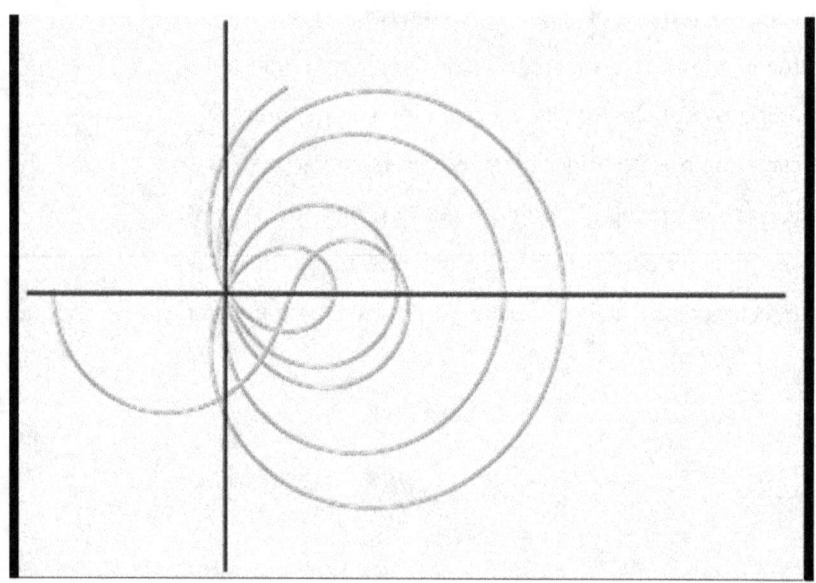

The above graph shows all nontrivial zeros as Origin intercepts when $\sigma = \dfrac{1}{2}$.

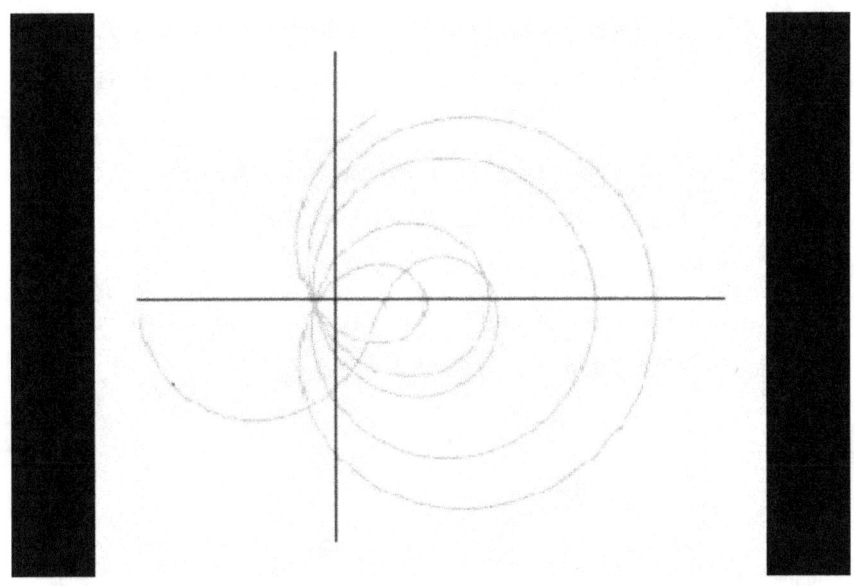

The above graph shows nil nontrivial zeros when $\sigma = \dfrac{2}{5}$.

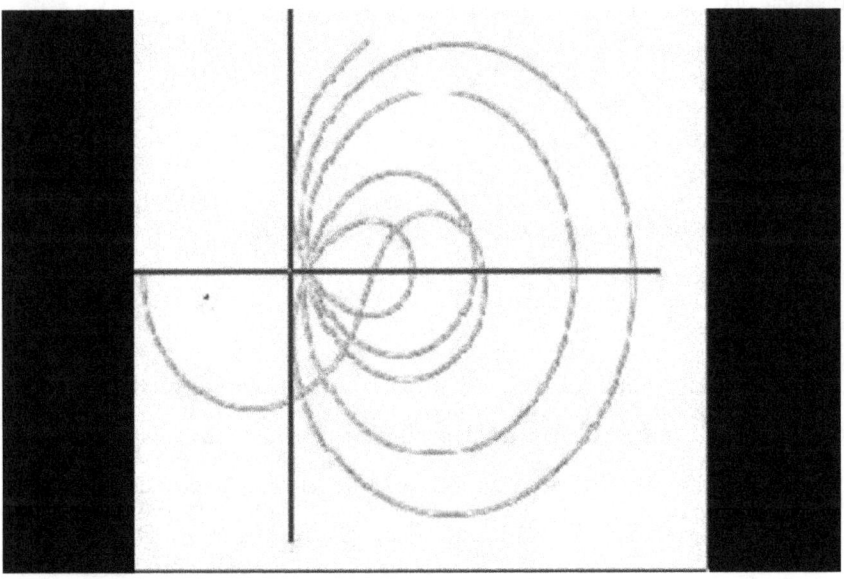

The above graph shows nil nontrivial zeros when $\sigma = \dfrac{3}{5}$.

We surmise from above three graphs on Riemann zeta function that when $\sigma = \dfrac{1}{2}$;

55

Gram[x=0,y=0] conjecture (Riemann hypothesis), Gram[y=0] conjecture, and Gram[x=0] conjecture can be combined as 'grand' Dirichlet-Gram-Riemann conjecture. When $\sigma \neq \frac{1}{2}$; Gram[x=0,y=0] (Riemann hypothesis) conjecture cannot exist and we only have 'virtual' Gram[y=0] conjecture associated with 'virtual' Gram[y=0] points and 'virtual' Gram[x=0] conjecture associated with 'virtual' Gram[x=0] points. 'Virtual' Gram[y=0] points and 'virtual' Gram[x=0] points have totally different values to their corresponding Gram[y=0] points and Gram[x=0] points.

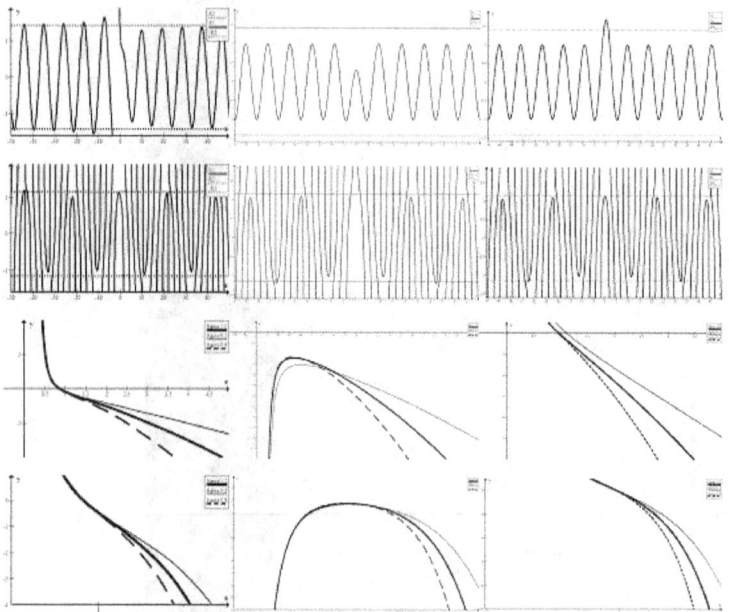

We depict the above computer-generated "combined" miniaturized version graphs from (Ting J. , Rigorous Proof for Riemann Hypothesis Using the Novel Sigma-power Laws and Concepts from the Hybrid Method of Integer Sequence Classification https://doi.org/10.5539/jmr.v8n3p9, 2016) and (Ting J. , Key Role of Dimensional Analysis Homogeneity in Proving Riemann Hypothesis and Providing Explanations on the Closely Related Gram Points https://doi.org/10.5539/jmr.v8n4p1, 2016) depicted using $\sigma = \frac{1}{5}$, $\sigma = \frac{1}{2}$, and $\sigma = \frac{4}{5}$ thus giving us a snapshot on "physical manifestations" of various Laws – taking into account the 'Errata Notice' mentioned in Chapter 1 on mathematical errors present in relevant equations. These graphs are derived out of Gram[x=0,y=0] conjecture (Riemann hypothesis), Gram[y=0] conjecture and Gram[x=0] conjecture. Respectively, the mentioned σ values are from Left to Right seen as three vertical columns of 'grouped' miniature graphs.

Table 1: **Completely Predictable problem of Even-odd number pairing.** Note the non-varying relationship between even and odd numbers.

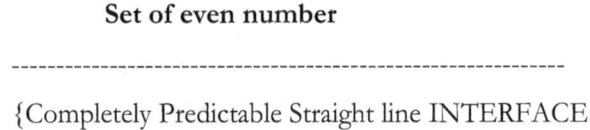

Set of even number

{Completely Predictable Straight line INTERFACE}

Set of odd number

Table 2: **Incompletely Predictable problem of Prime-composite number pairing.** Note the varying relationship between prime and composite numbers.

Set of prime number

vvvvvvvvvvvvvvvvvvvvvvvvvvvvvvvvvvvvvvvvvvvvvv

{Completely Predictable Jagged line INTERFACE}

Set of composite number

Table 3: **Completely Predictable problem on intercepts of y = sin x function.** Note the non-varying relationship between [solitary] 'Origin' intercept and x-axis intercepts.

Set of x-axis intercepts in y = sin x

{Completely Predictable Straight line INTERFACE}

Set of 'Origin' intercept in y = sin x

Table 4: **Incompletely Predictable problem on intercepts of Riemann zeta function (or its *proxy* Dirichlet eta function).** Note the varying relationship between all three types of Gram points.

Set of Gram[y=0] points ('usual' Gram points)

vvvvvvvvvvvvvvvvvvvvvvvvvvvvvvvvvvvvvvvvvvvvvv

{Completely Predictable Jagged line INTERFACE}

57

Set of Gram[x=0,y=0] points (nontrivial zeros)

vvvvvvvvvvvvvvvvvvvvvvvvvvvvvvvvvvvvvvvvvvv

{Completely Predictable Jagged line INTERFACE}

Set of Gram[x=0] points

Simple Containers: simple algorithms ("discrete") y = 2x & y = 2x - 1 and simple equations ("continuous") y = 2x, y = 2x - 1 & y = sin x. Complex Containers: complex algorithm ("discrete") Sieve of Eratosthenes and complex equation ("continuous") Riemann zeta function (or its *proxy* Dirichlet eta function). Completely Predictable problems: even-odd number pairing given by "discrete" y = 2x and y = 2x - 1, straight line pairing given by "continuous" y = 2x and y = 2x - 1, and axes intercepts given by "continuous" y = sin x. Incompletely Predictable problems: prime-composite number pairing given by "discrete" Sieve of Eratosthenes and axes intercepts given by "continuous" Riemann zeta function. In regards to 'Numerical relationship interface' and 'Axes intercept relationship interface', CFS of simple & complex properties are present infinitely often at both types of relationship interface for, respectively, Completely & Incompletely Predictable problems.

We can now provide a brief synopsis on Incompletely Predictable entities for Riemann zeta function. Gram points and virtual Gram points are dependently calculated using complex equation Riemann zeta function, ζ (s), or its proxy Dirichlet eta function, $\eta(s)$, in critical strip (denoted by $0 < \sigma < 1$). Gram[y=0], Gram[x=0] and Gram[x=0,y=0] points respectively refer to x-axis, y-axis and Origin intercepts at critical line (denoted by $\sigma = \frac{1}{2}$). Gram[y=0] and Gram[x=0,y=0] points are respectively synonymous with traditional 'Gram points' and nontrivial zeros. Virtual Gram[y=0] and virtual Gram[x=0] points respectively refer to x-axis and y-axis intercepts at non-critical lines (denoted by $\sigma \neq \frac{1}{2}$). Virtual Gram[x=0,y=0] points do not exist. Activities to prove associated open problem in number theory Riemann hypothesis and explain Gram[y=0] and Gram[x=0] points equate to solving Incompletely Predictable problems.

Solving our three open problems of Riemann hypothesis, Polignac's and Twin prime conjectures will involve obtaining and analyzing these Completely Predictable CFS of complex properties that are (intrinsically) present in complex equation & algorithm instead of hitting a **"mathematical impasse"** by analyzing CIS of Incompletely Predictable entities (extrinsically) generated by these equation & algorithm.

Obviously a lot of materials in Chapter 5 will belong to this chapter [and *vice versa*] due to "overlapping" presence of Completely Predictable and Incompletely Predictable entities in both 'Axes Intercept relationship Interface' and 'Numerical Relationship Interface'.

At 'Numerical relationship interface', uncountable complex numbers (**C**) include uncountable real numbers (**R**). **R** = countable rational numbers (**Q**) + uncountable irrational numbers (**R** − **Q**). **R** − **Q** = countable algebraic numbers + uncountable transcendental numbers. **Q** include countable integers (**Z**) which include countable whole numbers (**W**) which in turn include countable natural numbers (**N**). **N** is constituted by either countable even numbers (**E**) and countable odd numbers (**O**) or countable prime numbers (**P**), countable composite numbers (**C**) and Number '1'.

Then (i) Set **N** = Set **E** + Set **O**, (ii) Set **N** = Set **P** + Set **C** + Number '1', and (iii) Set **N** ⊂ Set **W** ⊂ Set **Z** ⊂ Set **Q** ⊂ Set **R** ⊂ Set **C**. With increasing magnitude, arbitrary Set **X** belongs to countable finite set (CFS), countable infinite set (CIS) or uncountable infinite set (UIS). Cardinality of Set **X**, $|\mathbf{X}|$, measures the "number of elements" in Set **X**. E.g. Set even **P** has CFS of even **P** with $|\text{even } \mathbf{P}| = 1$, Set **N** has CIS of **N** with $|\mathbf{N}| = \aleph 0$, and Set **R** has UIS of **R** with $|\mathbf{R}| = c$ (cardinality of the continuum).

In comparison at the 'Axes intercept relationship interface', uncountable Generated curves of Riemann zeta function include three types of Gram points viz. countable Gram[x=0,y=0] points, countable Gram[x=0] points & countable Gram[y=0] points. Then (i) Set all Gram points = Set Gram[x=0,y=0] points + Set Gram[x=0] points + Set Gram[y=0] points, and (ii) Set all Gram points ⊂ Set Generated curves. Note that Set negative Gram[y=0] point has CFS of negative Gram[y=0] point with $|$negative Gram[y=0] point$| = 1$.

Dependently calculated using complex algorithm Sieve of Eratosthenes, **P** and **C** are Incompletely Predictable numbers. Occurring over 2000 years ago (circa 300 BC), ancient Euclid's proof on infinitude of **P** in totality [viz. $|\mathbf{P}| = \aleph_0$ for Set **P**] predominantly by *reductio ad absurdum* (proof by contradiction) is earliest known but not the only proof for this simple problem in number theory.

Since then dozens of proofs have been devised such as three chronologically listed: Goldbach's Proof using Fermat numbers (written in a letter to Swiss mathematician Leonhard Euler, July 1730), Furstenberg's Topological Proof in 1955 (Furstenberg, 1955), and Filip Saidak's Proof in 2006 (Saidak, 2006). The strangest candidate is likely to be Furstenberg's Topological Proof.

In 2013, Yitang Zhang proved a landmark result showing some unknown even number 'N' < 70 million such that there are infinitely many pairs of P that differ by 'N' (Zhang, 2014). By optimizing Zhang's bound, subsequent Polymath Project collaborative efforts using a new refinement of GPY sieve in 2013 lowered 'N' to 246; and assuming Elliott-Halberstam conjecture and its generalized form have further lower 'N' to 12 and 6, respectively. Then 'N' has intuitively more than one valid values such that there are infinitely many pairs of P that differ by each of those 'N' values [thus proving existence of more than one Subset odd Pi with $|$ odd Pi $| = \aleph_0$]. We can only theoretically lower 'N' to 2 (in regards to P with 'small gaps') but there are still an infinite number of E prime gaps (in regards to P with 'large gaps') that require "the proof that each will generate its unique set of infinite P".

Two *overall* complex properties are perpetual presence of firstly, critical line location for Set **nontrivial zeros** of Riemann zeta function (indicating Riemann hypothesis to be true) and secondly, $|\mathbf{E}$ prime gaps$| = \aleph_0$ and Subsets odd **P** derived from associated **E** prime gaps all having $|$ odd **P** $| = \aleph_0$ (indicating Polignac's and Twin prime conjectures to be true).

Solving open problems of Riemann hypothesis, Polignac's and Twin prime conjectures involve deriving Complex Elementary Fundamental Laws-based solutions of Sigma-Power Laws for Riemann hypothesis, Plus Gap 2 Composite Number Continuous Law and Plus-Minus Gap 2 Composite Number Alternating Law for Polignac's and Twin prime conjectures obtained by (respectively) undertaking certain non-negotiable mathematical steps outlined by 'Riemann hypothesis mathematical foot-prints' (see Chapter 7 for further discussion) and 'Prime and Twin prime mathematical foot-prints' (see Chapter 8 for further discussion). We next provide collective statements for our two interfaces.

Collective statements on Axes intercept relationship interface:

For intercepts of y = sin x: Set of all axes intercepts = Set of x-axis intercepts + Single Origin intercept. Set of x-axis intercepts > Set of Origin intercept inequality forms a Completely Predictable "exact non-varying" Axes intercept relationship interface. Example of simple properties due to interactions between two independent sets of x-axis and Origin intercepts is the Completely Predictable timing of both intercepts eternal occurrences whereby they are all given by a simple algorithm: x-axis intercepts = $k\pi$, where k is an **Z** and $\pi = 3.14159$ [rounded off to five decimal places] is a transcendental number.

For intercepts of Riemann zeta function: Set of all axes intercepts = Set of x-axis intercepts + Set of y-axis intercepts + Set of Origin intercepts. Set of x-axis intercepts = Set of y-axis intercepts = Set of Origin intercepts equality forms an Incompletely Predictable "exact varying" Axes intercept relationship interface. Examples of complex properties due to interactions between two dependent sets of Gram[y=0] points and nontrivial zeros are Gram's Law & its violation, and Gram block which are Completely Predictable to periodically occur an infinite number of times, albeit in an Incompletely Predictable manner.

Collective statements on Numerical relationship interface:

For **E-O** number pairing: Set **N** = Set **E** + Set **O**. Set **E** = Set **O** equality forms a Completely Predictable "exact non-varying" Numerical relationship interface.

Example of simple properties due to interactions between two independent Set **E** and Set **O** are **E** gap = 2 and **O** gap = 2 whose eternal occurrences are Completely Predictable in timing.

For straight line pairing: [Set of two lines in combined length] = [Set of y = 2x line in length] + [Set of 2x - 1 line in length]. [Set of y = 2x line in length] = [Set of y = 2x - 1 line in length] forms a Completely Predictable "exact non-varying" Numerical relationship interface. Examples of simple properties due to interactions between two independent sets of y = 2x line and y = 2x - 1 line is their manifestation as two parallel infinite length straight lines perpetually separated apart by Completely Predictable *horizontal* distance = 0.5, *vertical* distance = 2, and *perpendicular* distance = √(0.25 - 0.0625) = √0.1875 = 0.433 [rounded off to three decimal places] -- this is an algebraic number.

For **P-C** number pairing: **N** = Set **P** + Set **C** + The number '1'. Set **C** > Set **P** inequality forms an Incompletely Predictable "exact varying" Numerical relationship interface. Examples of complex properties due to interactions between two dependent Set **P** and Set **C** relate to Composite Gap 2 Number appearances governed by Plus-Minus Gap 2 Composite Number Alternating Law and Plus Gap 2 Composite Number Continuous Law. These Completely Predictable laws are computationally applicable [albeit with Incompletely Predictable timing] on an eternal basis to designated **P** (generated from all even **P** gaps minus even **P** gap = 2 in the first law, and even **P** gap = 2 in the second law).

With same underlying complex equation, proof of Set Gram[x=0,y=0] points location on critical line should *dependently* confirm Set Gram[y=0] points & Set Gram[x=0] points location on this line. With same underlying complex algorithm, proof of ℵo cardinality value for Subsets odd **P** (from CIS **E** prime gaps = 2, 4, 6, 8, 10,...) constituting Set all **P** minus even **P** '2' should *dependently* confirm this cardinality value for Subset even **C** & Subset odd **C** (from CFS composite gaps = 1 & 2) constituting Set all **C**.

7 Solving Riemann hypothesis

This chapter refer to my research paper *Solving Incompletely Predictable problem Riemann hypothesis with Dirichlet Sigma-Power Law* which is fully outlined in Appendix 1.

Abstract for this paper:

Riemann hypothesis proposed all nontrivial zeros to be located on critical line of Riemann zeta function. Treated as Incompletely Predictable problem, we obtain Dirichlet Sigma-Power Law as final proof of solving this problem. This Law is derived as equation and inequation from original Dirichlet eta function (proxy function for Riemann zeta function). Performing a parallel procedure help explain closely related Gram points.

Mathematical Footprints as six identifiable steps to prove this hypothesis:

Step 1 Use $\eta(s)$, proxy for $\zeta(s)$, in critical strip.

Step 2 Apply Euler formula to $\eta(s)$.

Step 3 Obtain "simplified" Dirichlet eta function which intrinsically incorporates actual location [but not actual positions] of all nontrivial zeros#.

Step 4 Apply Riemann integral to "simplified" Dirichlet eta function in discrete (summation) format.

Step 5 Obtain Dirichlet Sigma-Power Law in continuous (integral) format as equation or inequation.

Step 6 Note exact and inexact DA homogeneity on their fractional exponents.

#Footnote: Respectively, Gram[y=0] points, Gram[x=0] points and Gram[x=0,y=0] (nontrivial zeros) are Incompletely Predictable entities with actual positions determined by setting $\sum Im\{\eta(s)\} = 0$, $\sum Re\{\eta(s)\} = 0$ and $\sum ReIm\{\eta(s)\} = 0$ to dependently calculate relevant positions of all preceding entities in neighborhood. Respectively, actual location of Gram[y=0] points, Gram[x=0] points and nontrivial zeros; and virtual Gram[y=0] points, virtual Gram[x=0] points and "absent" nontrivial zeros occur precisely at $\sigma = \dfrac{1}{2}$; and $\sigma \neq \dfrac{1}{2}$.

Outline of proof for Riemann hypothesis. To simultaneously satisfy two mutually inclusive conditions:

I. With rigid manifestation of exact DA homogeneity, Set nontrivial zeros with |nontrivial zeros| = $\aleph_0$ is located on critical line (viz. $\sigma = \dfrac{1}{2}$) when $2(1 - \sigma)$ [or $2(\sigma + 1)$] as $\sum$(all fractional exponents) = whole number '1' [or '3'] in Dirichlet Sigma-Power Law## as equation [or inequation].

II. With rigid manifestation of inexact DA homogeneity, Set nontrivial zeros with |nontrivial zeros| = $\aleph_0$ is not located on non-critical lines (viz. $\sigma \neq \dfrac{1}{2}$) when $2(1 - \sigma)$ [or $2(\sigma + 1)$] as $\sum$(all fractional exponents) = fractional number '$\neq$1' [or '$\neq$3'] in Dirichlet Sigma-Power Law## as equation [or inequation].

##Footnote: Derived from original $\eta(s)$ (proxy for $\zeta(s)$) as equation or inequation, this Law symbolizes end-result proof on Riemann hypothesis.

We now provide an outline of exact and inexact Dimensional analysis (DA) homogeneity. Respectively for 'base quantities' such as length, mass and time; their fundamental SI 'units of measurement' meter (m) is defined as distance travelled by light in vacuum for time interval 1/299 792 458 s with speed of light c = 299,792,458 ms^{-1}, kilogram (kg) is defined by taking fixed numerical value Planck constant h to be 6.626 070 15 X 10^{-34} Joules·second (Js) [whereby Js is equal to $kg m^2 s^{-1}$] and second (s) is defined in terms of

$\Delta vCs = \Delta(133Cs)h$ f s = 9,192,631,770 s^{-1}. Derived SI units such as J and ms^{-1} respectively represent 'base quantities' energy and velocity. The word 'dimension' is commonly used to indicate all those mentioned 'units of measurement' in well-defined equations.

DA is an analytic tool with DA homogeneity and non-homogeneity (respectively) denoting valid and invalid equation occurring when 'units of measurements' for 'base quantities' are "balanced" and "unbalanced" across both sides of the equation. E.g. equation 2 m + 3 m = 5 m is valid and equation 2 m + 3 kg = 5 mkg is invalid (respectively) manifesting DA homogeneity and non-homogeneity.

Let (2n) and (2n-1) be 'base quantities' in Dirichlet Sigma-Power Laws formatted in simplest forms as equations and inequations. E.g. DA on exponent $\frac{1}{2}$ in $(2n)^{\frac{1}{2}}$ in simplest form is correct but DA on exponent $\frac{1}{4}$ in equivalent $(2^2 n^2)^{\frac{1}{4}}$ *not* in simplest form is incorrect.

Fractional exponents as 'units of measurement' given by $(1 - \sigma)$ for equations and $(\sigma + 1)$ for inequations when $\sigma = \frac{1}{2}$ coincide with exact DA homogeneity###; and $(1 - \sigma)$ for equations and $(\sigma + 1)$ for inequations when $\sigma \neq \frac{1}{2}$ coincide with inexact DA homogeneity####.

###, ####Footnotes: Exact and inexact DA homogeneity occur in Dirichlet Sigma-Power Laws as equations or inequations for Gram[y=0] points, Gram[x=0] points and Gram[x=0,y=0] points (nontrivial zeros). Law of Continuity is a heuristic principle whatever succeed for the finite, also succeed for the infinite. Then these Laws which inherently manifest themselves on finite and infinite time scale should "succeed for the finite, also succeed for the infinite".

Respectively for equations and inequations, exact DA homogeneity at $\sigma = \frac{1}{2}$ denotes

66

$\sum$(all fractional exponents) as 2(1 − σ) and 2(σ + 1) equates to ["exact"] whole number '1' and '3'; and inexact DA homogeneity at σ ≠ $\frac{1}{2}$ denotes $\sum$(all fractional exponents) as 2(1−σ) and 2(σ +1) equates to ["inexact"] fractional number '≠1' and '≠3'.

The above preliminary materials should now put readers in a strong position to understand my research paper. The main body of my research paper – readers to refer to this area as provided in Appendix 1 – then outline using correct and complete mathematical arguments how rigorous proof for Riemann hypothesis and explaining the closely related Gram points are derived. These correct and complete mathematical arguments will not currently be elaborated upon in this book.

Rigorous proof for Riemann hypothesis is summarized as Theorem Riemann I – IV below [Note: QED = *quod erat demonstrandum*]:

As preliminary, we supply the following important mathematical arguments.

For 0 < σ < 1, then 0 < 2(1−σ) < 2. The only whole number between 0 and 2 is '1' which coincide with σ = $\frac{1}{2}$. When 0 < σ < $\frac{1}{2}$ and $\frac{1}{2}$ < σ < 1, then 0 < 2(1−σ) < 1 and 1 < 2(1−σ) < 2.

For 0 < σ < 1, 2 < 2(σ + 1) < 4. The only whole number between 2 and 4 is '3' which coincide with σ = $\frac{1}{2}$. When 0 < σ < $\frac{1}{2}$ and $\frac{1}{2}$ < σ < 1, then 2 < 2(σ +1) < 3 and 3 < 2(σ +1) < 4.

Legend: R = all real numbers. For 0<σ<1, σ consist of 0<R<1. For 0 < 2(1−σ) < 2 and 2 < 2(σ +1) < 4, 2(1−σ) and 2(σ +1) must (respectively) consist of 0 < R < 2 and 2 < R < 4.

An important caveat is that previously used phrases such as "fractional exponent σ" and "$\sum$(all fractional exponents) = whole number '1' [or '3'] and fractional number '≠1' [or '≠3']", although not incorrect per se, should respectively be replaced by "real number exponent σ" and "$\sum$(all real number exponents) = whole number '1' [or '3']

67

and real number '≠1' [or '≠3']#####" for complete accuracy. We apply this caveat to Theorem Riemann I – IV.

#####Footnote: As whole numbers ⊂ real numbers, one could also depict this phrase as "∑(all real number exponents) = real number '1' [or '3'] and real number '≠1' [or '≠3']".

Theorem Riemann I. Derived from proxy Dirichlet eta function, "simplified" Dirichlet eta function will exclusively contain de novo property for actual location [but not actual positions] of all nontrivial zeros.

Proof. The phrase "actual location [but not actual positions] of all nontrivial zeros" can be validly shortened to "actual location of all nontrivial zeros" as used in Theorem Riemann II, III and IV. The proof for Theorem Riemann I is now complete as it successfully incorporates proof for Lemma 3.1 QED.

Theorem Riemann II. Dirichlet Sigma-Power Law [in continuous (integral) format] as equation and inequation which are both derived from "simplified" Dirichlet eta function [in discrete (summation) format] will exclusively manifest exact DA homogeneity in equation and inequation only when real number exponent $\sigma = \frac{1}{2}$.

Proof. The proof for Theorem Riemann II is now complete as it successfully incorporates proofs from Proposition 3.2 on derivation for equation and inequation of Dirichlet Sigma-Power Law [with both containing de novo property for "actual location of all nontrivial zeros"] and Proposition 3.3 on manifestation of exact DA homogeneity in Dirichlet Sigma-Power Law as equation and inequation when real number exponent $\sigma = \frac{1}{2}$ QED.

Theorem Riemann III. Real number exponent $\sigma = \frac{1}{2}$ in Dirichlet Sigma-Power Law as equation and inequation satisfying exact DA homogeneity is identical to σ variable in Riemann hypothesis which propose σ to also have exclusive value of $\frac{1}{2}$ (representing critical line) for "actual location of all nontrivial zeros", thus fully supporting Riemann hypothesis to be true with further clarification by Theorem Riemann IV.

Proof. Since s = σ ± ıt, complete set of nontrivial zeros which is defined by η(s) = 0 is exclusively associated with one (and only one) particular η(σ ± ıt) = 0 value solution, and by default one (and only one) particular σ [conjecturally] = $\frac{1}{2}$ solution. When performing exact DA homogeneity on Dirichlet Sigma-Power Law as equation and inequation [with both containing de novo property for "actual location of all nontrivial zeros"], the phrase "If real number exponent σ has exclusively $\frac{1}{2}$ value, only then will exact DA homogeneity be satisfied" implies one (and only one) possible mathematical solution. Theorem Riemann III reflect Theorem Riemann II on presence of exact DA homogeneity for σ = $\frac{1}{2}$ in Dirichlet Sigma-Power Law as equation and inequation. This Law has identical σ variable as that referred to by Riemann hypothesis [whereby σ here uniquely refer to critical line]. The proof for Theorem Riemann III is now complete as it independently refers to simultaneous association of confirmed (i) solitary σ = $\frac{1}{2}$ value in Dirichlet Sigma-Power Law as equation and inequation satisfying exact DA homogeneity and (ii) critical line defined by solitary σ = $\frac{1}{2}$ value being the "actual location [but with no request to determine actual positions]" of all nontrivial zeros as proposed in original Riemann hypothesis QED.

Theorem Riemann IV. Condition 1. All σ ≠ $\frac{1}{2}$ values (non-critical lines), viz. 0 < σ < $\frac{1}{2}$ and $\frac{1}{2}$ < σ < 1 values, exclusively does not contain "actual location of all nontrivial zeros" [manifesting de novo inexact DA homogeneity in equation and inequation], together with Condition 2. One (and only one) σ = $\frac{1}{2}$ value (critical line) exclusively contains "actual location of all nontrivial zeros" [manifesting de novo exact

69

DA homogeneity in equation and inequation], fully support Riemann hypothesis to be true when these two mutually inclusive conditions are met.

Proof. Condition 2 Theorem Riemann IV simply reflect proof from Theorem Riemann III [incorporating Proposition 3.3] for "actual location of all nontrivial zeros" exclusively on critical line manifesting de novo exact DA homogeneity $\sum$(all real number exponents) = whole number '1' for equation [or '3' for inequation]. The proof for Condition 2 Theorem Riemann IV is now complete QED. Corollary 3.4 confirms de novo inexact DA homogeneity manifested as $\sum$(all real number exponents) = real number '$\neq$1' for equation [or '$\neq$3' for inequation] by all $\sigma \neq \frac{1}{2}$ values (non-critical lines) that are exclusively not associated with "actual location of all nontrivial zeros". Applying inclusion-exclusion principle: Exclusive presence of nontrivial zeros on critical line for Condition 2 Theorem Riemann IV implies exclusive absence of nontrivial zeros on non-critical lines for Condition 1 Theorem Riemann IV. The proof for Condition 1 Theorem Riemann IV is now complete QED.

The outline of conclusion section of my research paper is provided next.

In our Hybrid method of Integer Sequence classification, a formula is either non-Hybrid or Hybrid integer sequence. Inequation with two 'necessary' Ratio (R) or equation with one 'unnecessary' R contains non-Hybrid integer sequence. Equation with one 'necessary' R contains Hybrid integer sequence. "In the limit" Hybrid integer sequence approach unique Position X, it becomes non-Hybrid integer sequence for all Positions $\geq$ Position X.

Consider kinetic energy (KE) in MJ with m_o = rest mass in kg and v = velocity in m s^{-1}. In classical mechanics concerning low velocity with v << c, Newtonian KE = $\frac{1}{2}m_o v^2$. In relativistic mechanics concerning high velocity with v $\geq$ 0.01c, Relativistic

$$KE = \frac{m_o c^2}{\sqrt{(1-(v^2/c^2))}} - m_o c^2$$. Obtained from the later by binomial approximation or by taking first two terms of Taylor expansion for reciprocal square root, the former approximates the later well at low speed.

We arbitrarily divide DA homogeneity into inexact DA homogeneity for ["<100% accuracy"] Newtonian KE and exact DA homogeneity for ["100% accuracy"] Relativistic KE. "In the limit" ['<100% accuracy'] Newtonian KE at low speed approach ['100% accuracy'] Relativistic KE at high speed, we achieve perfection.

Analogy: "In the limit" all three version of Dirichlet Sigma-Power Laws for Gram[y=0] points, Gram[x=0] points and nontrivial zeros as '<100% accuracy' inequations approach perfection as '100% accuracy' equations, compliance with inexact DA homogeneity becomes compliance with exact DA homogeneity. We note R1 terms in all inequations contain (2n) and (2n-1) 'base quantities' but these are not endowed with fractional exponent (σ+1) as relevant 'unit of measurement'. As Incompletely Predictable problems, we gave relatively elementary proof of Riemann hypothesis and explain closely related Gram points whereby various "meta-properties" such as exact and inexact DA homogeneity occur in (respectively) equations and inequations of relevant Dirichlet Sigma-Power Laws. Harnessed key benefit from successful proof for Riemann hypothesis is often stated as "With this one solution, we have proven five hundred theorems or more at once". This apply to important theorems in number theory that rely on properties of Riemann zeta function or Dirichlet eta function such as location of trivial and nontrivial zeros. E.g., we delineate prime number theorem by prime counting function $\pi(x)$ [which is defined as number of primes $\leq$ x].

This chapter refer to my research paper *Solving Incompletely Predictable problems Polignac's and Twin Prime conjectures using Information-Complexity conservation* which is fully outlined in Appendix 2.

Abstract for this paper:

Prime numbers are Incompletely Predictable numbers calculated using complex algorithm Sieve of Eratosthenes. Involving proposals that prime gaps and associated sets of prime numbers are infinite in magnitude, Twin prime conjecture deals with even prime gap 2 and is a subset of Polignac's conjecture which deals with all even prime gaps 2, 4, 6, 8, 10,.... Treated as Incompletely Predictable problems, we solve these conjectures with research method Information-Complexity conservation to get Plus Gap 2 Composite

Number Continuous Law and Plus-Minus Gap 2 Composite Number Alternating Law.

Mathematical Footprints as six identifiable steps to prove these conjectures:

Step 1 Considering x ∈ N, obtain Dimensions $(2x-2)^1$, $(2x-4)^1$, $(2x-5)^1$, $(2x-7)^1$, $(2x-8)^1$, $(2x-9)^1$, ..., $(2x-\infty)^1$ with specific groupings to constitute all elements of Set P [culminating in obtaining all prime gaps (= E prime gaps + Solitary O prime gap) with |all prime gaps| = $\aleph_0$]. Note Dimension $(2x-2)^1$ represents x = 1 (Number '1') which is neither P nor C. Step 2 Considering i ∈ E, confirm perpetual recurrences of individual E prime gap = i (associated with its unique odd Pi) occur only when depicted as specific groupings of these Dimensions endowed with exponent '1' for all ranges of x.

Step 3 Perform DA on exponent '1' in these Dimensions.

Step 4 Perform DA on equation Set odd P = $\sum_{i=2}^{\infty}$ *Subset odd Pi* to obtain |odd P|

= |odd Pi| = ℵo whereby Subset odd Pi is derived from its associated unique E prime gap = i with |E prime gaps| = ℵo.

Step 5 Confirm 'Prime number' variable and 'Prime gap' variable complex algorithm "containing" all P with knowing their overall actual location [but not actual positions]#.

Step 6 Derive Plus-Minus Gap 2 Composite Number Alternating Law and Plus Gap 2 Composite Number Continuous Law using Information-Complexity conservation.

#Footnote: This phrase implies all P (and C) are treated as Incompletely Predictable numbers. Actual positions will require using complex algorithm Sieve of Eratosthenes to dependently calculate positions of all preceding P (and C) in the neighborhood.

Outline of proof for Polignac's and Twin prime conjectures. Requires simultaneously satisfying two mutually inclusive conditions:

I. With rigid manifestation of DA homogeneity, quantitive## fulfillment by considering i ∈ E for each Subset odd Pi generated by E prime gap = i from Set E prime gaps occurs only if solitary cardinality value is present in equation Set odd P =

$$\sum_{i=2}^{\infty} Subset\ odd\ Pi$$

with |odd P| = |odd Pi| = |E prime gaps| = ℵo.

II. With rigid manifestation of DA non-homogeneity, quantitive## fulfillment by considering i ∈ E for each Subset odd Pi generated by E prime gap = i from Set E prime gaps does not occur if more than one cardinality values are present in equation

Set odd P > $$\sum_{i=2}^{\infty} Subset\ odd\ Pi$$ with |E prime gaps| = ℵo having incorrect

|Subset(s) odd P| = N (finite value) and/or Set odd P > $$\sum_{i=2}^{N} Subset\ odd\ Pi$$ with |odd Pi| = ℵo having incorrect |E prime gaps| = N (finite value).

##Footnote: Qualitative fulfillment of |odd P| = |odd Pi| = |all E prime gaps| =

ℵo equates to Plus-Minus Gap 2 Composite Number Alternating Law being precisely obeyed by all E prime gaps apart from first E prime gap precisely obeying Plus Gap 2 Composite Number Continuous Law. Derived using Information-Complexity conservation, these Laws symbolize "end-result" proof on Polignac's and Twin prime conjectures. Law of Continuity is a heuristic principle whatever succeed for the finite, also succeed for the infinite. Then these Laws which inherently manifest 'Gap 2 Composite Number' on finite and infinite time scale should in principle "succeed for the finite, also succeed for the infinite".

We now provide an outline on Dimensional analysis (DA) on Cardinality and "Dimensions". For 'base quantities' such as length, mass and time; their fundamental SI 'units of measurement' are [respectively] given by meter (m), kilogram (kg) and second (s). The word 'dimension' is commonly used to denote 'units of measurement' in well-defined equations.

DA is an analytic tool with resulting DA homogeneity and non-homogeneity (respectively) denoting valid and invalid equation when 'units of measurements' are "balanced" and "unbalanced" across both sides of the equation. E.g. 2 m + 3 m = 5 m is a valid equation but 2 m + 3 kg = 5 mkg is an invalid equation.

We use "Dimensions" to denote well-defined Incompletely Predictable entities obtained from Information-Complexity conservation. Relevant "Dimensions" dependently represent Number '1', P and C. Then by default any (sub)sets of P and C in well-defined equations can also be represented by their corresponding "Dimensions". We can apply Dimensional analysis to "Dimensions" from Information-Complexity conservation and cardinality of relevant sets in certain well-defined equations.

Let X denote E, O, N [which are classified as Completely Predictable numbers], P and

74

C [which are classified as Incompletely Predictable numbers]. For x = 1, 2, 3, 4, 5,..., ∞; consider all X ≤ x. Then this "all X ≤ x" is definition for X-π(x) [denoting "X counting function"] resulting in following two types of equations coined as (I) 'Exact' equation Nπ(x) = E-π(x) + O-π(x) with "non-varying" relationships E-π(x) = O-π(x) for all x = E and E-π(x) = O-π(x) - 1 for all x = O, and (II) 'Inexact' equation N-π(x) = 1 + P-π(x) + C-π(x) with "varying" relationships P-π(x) > C-π(x) for all x ≤ 8; P-π(x) = C-π(x) for x = 9, 11, and 13; and P-π(x) < C-π(x) for x = 10, 12, and all x ≥ 14.

Let "Dimensions" and different (sub)sets of E, O, N, P and C be 'base quantities'. Then exponent '1' of "Dimensions" and cardinality of these (sub)sets in well-defined equations Polignac's and Twin Prime conjectures are corresponding 'units of measurement'. Performing DA on 'Dimensions" for PC pairing are depicted later on [in my paper]. Performing DA on cardinality are depicted next.

For Set N = Set E + Set O, then |N| = |E| + |O| => ℵo = ℵo + ℵo thus conforming with DA homogeneity.
For Set N = Set P + Set C + Number '1', then Set N - Number '1' = Set P + Set C and |N - Number '1'| = |P| + |C| => ℵo = ℵo + ℵo thus conforming with DA homogeneity.
For Set N - Set even P - Number '1'= Set odd P + Set even C + Set odd C, then |N- even P - Number '1'| = |odd P| + |even C| + |odd C| => ℵo = ℵo + ℵo + ℵo thus conforming with DA homogeneity.

Symbolically represented by all available O prime gap = 1 and E prime gaps = 2, 4, 6, 8, 10,...; O composite gap = 1 and E composite gap = 2; and O natural gap = 1; then |Gap 1 N - Gap 1 P - Number '1'| = |Gap 2 P| + |Gap 4 P| + |Gap 6 P| + |Gap 8 P| + |Gap 10 P| + ... + |Gap 1 C| + |Gap 2 C| => ℵo = ℵo + ℵo + ℵo + ℵo + ℵo + ... ℵo + ℵo thus conforming with DA homogeneity. It is known that |Gap 1 P| = |Number '1'| = 1 and |Gap 1 N| = |Gap 1 C| = |Gap 2 C| = ℵo. Then solving

75

Polignac's & Twin prime conjectures translate to successfully proving $|\text{Gap 2 P}|$ = $|\text{Gap 4 P}|$ = $|\text{Gap 6 P}|$ = $|\text{Gap 8 P}|$ = $|\text{Gap 10 P}|$ = ... = $\aleph_0$ with $|\text{E prime gaps}|$ = $\aleph_0$.

The above preliminary materials should now put readers in a strong position to understand my research paper. The main body of my research paper – readers to refer to this area as provided in Appendix 2 – then outline using the correct and complete mathematical arguments on how rigorous proofs for Polignac's and Twin prime conjectures are derived. These mathematical arguments will not be further elaborated upon in this book.

Rigorous proofs for Polignac's and Twin prime conjectures are summarized as Theorem Polignac-Twin prime I – IV below [Note: QED = quod erat demonstrandum]:

Theorem Polignac-Twin prime I. Incompletely Predictable prime numbers Pn = 2, 3, 5, 7, 11, ..., ∞ or composite numbers Cn = 4, 6, 8, 9, 10, ..., ∞ are CIS with overall actual location [but not actual positions] of all prime or composite numbers accurately represented by complex algorithm involving prime gaps GPi viz. $P_{n+1} = 2 + \sum_{i=1}^{n} GPi$ or involving composite gaps GCi viz. $C_{n+1} = 4 + \sum_{i=1}^{n} GCi$ whereby prime & composite numbers are symbolically represented here with aid of 'n' notation instead of usual 'i' notation; and i & n = 1, 2, 3, 4, 5, ..., ∞. Number '2' in first algorithm represents P1, the very first (and only even) P. Number '4' in second algorithm represent C1, the very first (and even) C.

Proof. We treat above algorithms as unique mathematical objects looking for key intrinsic properties and behaviors. Each P or C is assigned a unique prime or composite gap. Absolute number of P or C and (thus) prime or composite gaps are infinite in magnitude. As original formulae containing all P or C by themselves (viz.

76

without supplying prime or composite gaps as "input information" to generate P or C as "output complexity"), these algorithms intrinsically incorporate overall actual location [but not actual positions] of all P or C. The proof is now complete for Theorem Polignac-Twin prime I QED.

Theorem Polignac-Twin prime II. Set of prime gaps GPi = 2, 4, 6, 8, 10, ..., ∞ is infinite in magnitude whereby these prime gaps accurately and completely represented by Dimensions $(2x - 7)^1$, $(2x - 8)^1$, $(2x - 9)^1$, ..., $(2x - \infty)^1$ must satisfy Information-Complexity conservation in a consistent manner.

Proof. Part I of Proposition 4.2 proved all P are represented by Dimension $(2x - N)^1$ with N $\geq$ 7 for any given x value (except for x = 2 & 3 values). Note that although x = 1 is neither P nor C, it is validly represented by Dimension $(2x - 2)^1$. If each P is endowed with a specific prime gap value, then each such prime gap must [via logical mathematical deduction] be represented by Dimension $(2x - N)^1$. We advocate this nominated method of prime gap representation using Dimensions be [purportedly] the only way to achieve Information-Complexity conservation. The preceding mathematical statements are correct as there is a unique prime gap value associated with each P. Proposition 5.1 below based on principles from Set theory provides further supporting materials that prime gaps are infinite in magnitude. The proof is now complete for Theorem Polignac-Twin prime II QED.

Theorem Polignac-Twin prime III. To maintain Dimensional analysis (DA) homogeneity, those Dimensions $(2x - N)^1$ from Theorem Polignac-Twin prime II must contain eternal repetitions of well-ordered sets constituted by Dimensions $(2x - 7)^1$, $(2x - 8)^1$, $(2x - 9)^1$, $(2x - 10)^1$, $(2x - 11)^1$, ..., $(2x - \infty)^1$.

Proof. This Theorem is stated in greater details as "To maintain DA homogeneity, those aforementioned [endowed with exponent 1] Dimensions $(2x - N)^1$ from Theorem Polignac-Twin prime II must repeat themselves indefinitely in following specific combinations – (i) Dimension $(2x - 7)^1$ only appearing as twin [two-times-in-a-row] and quadruplet [fourtimes-in-a-row] sequences, and (ii) Dimensions $(2x - 8)^1$, $(2x - 9)^1$, $(2x - 10)^1$, $(2x - 11)^1$, ..., $(2x - \infty)^1$ appearing as progressive groupings of E 2, 4, 6, 8, 10,..., ∞." To accommodate the only even P '2', exceptions to this DA homogeneity compliance will expectedly occur right at beginning of P sequence – (i) one-off appearance of Dimensions $(2x - 2)^1$, $(2x - 4)^1$ and $(2x - 5)^1$ and (ii) one-off appearance of Dimension $(2x - 7)^1$ as a quintuplet [five-times-in-a-row] sequence which is equivalent to (eternal) non-appearance of Dimension $(2x - 6)^1$ at x = 4. [We again note Dimension $(2x - 2)^1$ validly represent Number '1' which is neither P nor C.] These sequentially arranged sets are CFS whereby from x = 11 onwards, each set always commence initially as 'baseline' Dimension $(2x - 7)^1$ at x = O values and always end with its last Dimension at x = E values. Each set also have varying cardinality with values derived from all E; and correctly combined sets always intrinsically generate two infinite sets of P and, by default, C in an integrated manner. Our Theorem Polignac-Twin prime

III simply represent a mathematical summary derived from Section 3 & 4 of all expressed characteristics of Dimension $(2x - N)^1$ when used to represent P with intrinsic display of DA homogeneity. See Proposition 5.2 below for further details on DA aspect. The proof is now complete for Theorem Polignac-Twin prime III QED.

Theorem Polignac-Twin prime IV. Aspect 1. The "quantitive" aspect to existence of both prime gaps and their associated prime numbers as sets of infinite magnitude will be shown to be correct by utilizing principles from Set theory. Aspect 2. The "qualitative" aspect to existence of both prime gaps and their associated prime numbers as sets of infinite magnitude will be shown to be correct by 'Plus-Minus Gap 2 Composite Number Alternating Law' and 'Plus Gap 2 Composite Number Continuous Law'.

Proof. Required concepts from Set theory involve cardinality of a set with its 'well ordering principle' application. Supporting materials for these concepts based on 'pigeonhole principle' in relation to Aspect 1 are outlined in Proposition 5.1 below. 'Plus-Minus Gap 2 Composite Number Alternating Law' is applicable to all E prime gaps [apart from first E prime gap = 2 for twin primes]. The prime gap = 2 situation will obey 'Plus Gap 2 Composite Number Continuous Law'. These Laws are essentially Laws of Continuity inferring underlying intrinsic driving mechanisms that enables infinity magnitude association for both prime gaps & prime numbers to co-exist. By the same token, these Laws have the important implication that they must be applicable to those relevant prime gaps on an perpetual time scale. Supporting materials in relation to Aspect 2 are found in Proposition 4.2 above. The proof is now complete for Theorem Polignac-Twin prime IV QED.

We note two mutually inclusive conditions: Condition 1. Presence of all Dimensions that repeat themselves on an indefinite basis and with exponent of '1' will give rise to complete sets of P & C ["DA-wise one & only one mathematical possibility argument" associated with inevitable de novo DA homogeneity], and Condition 2. Presence of any Dimension(s) that do not repeat itself (themselves) on an indefinite basis or with exponent other than '1' will give rise to incomplete set of P & C or incorrect set of non-P & non-C ["DA-wise mathematical impossibility argument" associated with inevitable de novo DA non-homogeneity].

When met, these two conditions will fully support the point that CFS Dimensions representations of P & C [with respective prime & composite gaps] are totally accurate. Condition 1 reflect proof from Theorem Polignac-Twin prime III above as all P & C are associated with DA homogeneity when their Dimensions are endowed with exponent of '1'. Condition 2 invoke corollary on inevitable appearance of incomplete P or C or non-P or non-C [associated with DA non-homogeneity] being tightly incorporated into this mathematical framework. See Propositions 5.1 and 5.2, and Corollary 5.3 below for supporting materials on DA homogeneity & non-homogeneity.

79

We analyze P (& C) in terms of (i) measurements based on cardinality of CIS and (ii) pigeonhole principle which states that if n items are put into m containers, with n>m, then at least one container must contain more than one item. We note that ordinality of all infinite P (& C) is "fixed" implying that each one of the infinite well-ordered Dimension sets conforming to CFS type as constituted by Dimensions $(2x-7)^1$, $(2x-8)^1$, $(2x-9)^1$, $(2x-10)^1$, $(2x-11)^1$, ..., $(2x-\infty)^1 1$ on respective gaps for P (& C) must also be "fixed".

The outline of conclusion section of my research paper is provided next.

The harnessed property CIS of [Completely Predictable] natural numbers 1, 2, 3, 4, 5, 6, 7,... having CIS of [Completely Predictable] natural gaps 1, 1, 1, 1, 1, 1,... are constituted by three dependent sets of numbers:

(i)   CIS of [Incompletely Predictable] odd prime numbers 3, 5, 7, 11, 13, 17,... having CIS of [Incompletely Predictable] prime gaps 2, 2, 4, 2, 4,... plus CFS of solitary [Incompletely Predictable] even prime number 2 having CFS of [Incompletely Predictable] prime gap 1,

(ii)   CIS of [Incompletely Predictable] even and odd composite numbers 4, 6, 8, 9, 10, 12,... having CIS of [Incompletely Predictable] composite gaps 2, 2, 1, 1, 2, 2,.... and

(iii)   CFS of solitary odd number '1' [neither prime nor composite]. Treated as Incompletely Predictable problems endowed with "meta-properties", we gave relatively elementary proofs on Polignac's and Twin prime conjectures based on this harnessed property by performing Dimensional analysis on (sub)sets and "Dimensions" of prime and composite numbers, and obtaining 'Plus-Minus Gap 2 Composite Number Alternating Law' and 'Plus Gap 2 Composite Number Continuous Law'.

Prime number theorem describes asymptotic distribution of prime numbers among positive integers by formalizing intuitive idea that prime numbers become less common as they become larger through precisely quantifying rate at which this occurs using probability. Nontrivial zeros [from 'Axes intercept relationship interface' relevant to Riemann hypothesis] and prime numbers [from 'Numerical relationship interface' relevant to prime number theorem] are Incompletely Predictable entities and numbers.

Deep-seated connections exist between Riemann hypothesis and prime number theorem (which is fully delineated by prime counting function [denoted here with $\pi(x)$]). Solving Incompletely Predictable problem Riemann hypothesis is instrumental in proving efficacy of techniques that estimate $\pi(x)$ efficiently. This should now confirm "best possible" bound for error ("smallest possible" error) of prime number theorem.

In mathematics, logarithmic integral function or integral logarithm li(x) is a special function. Relevant to problems of physics and with number theoretic significance, it occurs in prime number theorem as an estimate of $\pi(x)$ whereby the form of this special function is defined so that li(2) = 0; viz. $\mathrm{li}(x) \equiv \int_{2}^{x} \dfrac{du}{\ln u}$ = li(x) - li(2). There are less accurate ways of estimating $\pi(x)$ such as conjectured by Gauss and Legendre at end of 18th century. This is approximately x/ln x in the sense $\lim\limits_{x\to\infty} \left(\dfrac{\pi(x)}{nx/\ln x} \right) = 1$.

Skewes' number is any of several extremely large numbers used by South African mathematician Stanley Skewes as upper bounds for smallest natural number x for which li(x)<$\pi(x)$. These bounds have since been improved by others: there is a crossing near $e^{727.95133}$ but it is not known whether this is the smallest. John Edensor Littlewood, who was Skewes' research supervisor, proved in 1914 that there is such a [first] number; and found that sign of difference $\pi(x)$ - li(x) changes infinitely often. This refute all prior numerical evidence that seem to suggest li(x) was always more than

81

$\pi(x)$.

The key point is [100% accurate] $\pi(x)$ mathematical tool being "wrapped around" by [less-than-100% accurate] approximate mathematical tool li(x) infinitely often via this 'sign of difference' changes meant that li(x) is the most efficient approximate mathematical tool. Contrast this with "crude" x/ln x approximate mathematical tool where values obtained diverge away from $\pi(x)$ at increasingly greater rate when larger range of prime numbers are studied.

9 Monstrous Moonshine (plus Medical statistics & acid-base physiology)

By an L-function, we generally refer to a Dirichlet series with a functional equation and an Euler product. Contextually, the simplest example of an L-function is Riemann zeta function on which the 1859 Riemann hypothesis is based upon. L-functions are ubiquitous in number theory and hence have applications to mathematical physics and cryptography. They arise from and encode information about a number of mathematical objects and it is necessary to exhibit these objects along with the L-functions themselves since typically we need these objects to compute L-functions.

For examples, L-functions can come from modular forms, elliptic curves, number fields, and Dirichlet characters, as well as more generally from automorphic forms, algebraic varieties, and Artin representations. Broadly based on these examples, the mammoth 'L-functions and Modular Forms Database' (LMFDB) creation was conducted with massive team-effort collaboration from an international group of more than 80 researchers from 12 countries which included prominent mathematicians such as from American Institute of Mathematics in United States, University of Bristol in United Kingdom, and Dartmouth College in United States.

The LMFDB idea was first conceived at an American Institute of Mathematics workshop in 2007. Six years after commencing the LMFDB project [website address http://www.lmfdb.org], its launching was celebrated on May 10, 2016. In effect, LMFDB can be considered an uncharted mathematical terrain providing a detailed atlas of mathematical objects that highlights deep relationships and serves as a guide to latest research happening in physics, computer science and mathematics. Elliptic curves arise naturally in many parts of mathematics and can be described by a simple cubic equation. They also form the basis of cryptographic protocols used by most of the major internet companies including Google, Facebook and Amazon. Modular forms are more mysterious objects constituted by complex functions with an almost unbelievable degree of symmetry.

The two mathematical worlds of elliptic curves and modular forms are remarkably connected via their L-functions. It is this deep connection that was in essence required in the late 20th century by famous British number theorist Andrew John Wiles to successfully achieve his proof of Fermat's Last Theorem. To put into perspective the importance of LMFDB in relation to active research areas such as involving Monstrous moonshine (Moonshine theory), Mathieu moonshine, and Umbral moonshine with their conjectured roles in Quantum gravity and String theory; we think that most physicists would have a positive opinion or consensus on the potential role of these research areas in successfully merging gravity with Grand Unified Theory (GUT) – consisting of the unification of electromagnetism, weak nuclear force, and strong nuclear force – thus giving rise to the holy grail Theory of Everything (TOE).

We briefly divert here to mention that the name 'Standard Model of particle physics', commenced in the 1970s, denotes the theory describing three of the four known fundamental forces in the universe (viz. the electromagnetic, weak, and strong interactions of GUT), as well as classifying all known elementary particles. Despite all its predictive power, it is not 'perfect" in that it can't explain gravity, dark matter or dark energy.

String theories assume that fundamental building blocks of the universe are strings instead of point particles. String duality is a class of symmetries in physics that link different String theories, with K3 surfaces appearing almost ubiquitously in string duality. A K3 surface is a complex or algebraic smooth minimal complete surface that is regular and has trivial canonical bundle. Not least because of this difficulty of multiple String theories (and hence multiple possibilities), an alternative view is that all four fundamental forces of nature will always exist as the current *status quo* with gravity obeying laws [perhaps endowed with certain "continuous" Completely Predictable properties] derived from Einstein's Theory of General Relativity and the three forces of GUT obeying laws [perhaps endowed with certain "discrete" Incompletely Predictable properties] based on Quantum mechanics. Alternatively stated, nature will intrinsically never allow the mathematical merging together of those two totally incompatible situations; namely, the "continuous" property on the one hand and "discrete" property on the other hand. Despite this issue, LMFDB with one of its crucial features acting as "intricate catalog of mathematical objects" will, metaphorically speaking, be the source supplying the required mathematical objects in those mentioned research areas.

In the grand scheme of things, this paper manifests the classically encountered phenomenon that pure and applied mathematics during, and resulting from, derivation of many mathematical proofs are largely inseparable. Some of the less conventional aspects of resulting applied mathematics in regards to the following (depicted from biologist-to-physicist point of view with highest-to-lowest decreasing hierarchical order) are intuitively useful:

I. Living Things obeying Complex Emergent Fundamental Laws
II. Living Things obeying Simple Emergent Fundamental Laws
III. Nonliving Things obeying Complex Elementary Fundamental Laws
IV. Nonliving Things obeying Simple Elementary Fundamental Laws

In this context, our Incompletely Predictable problems of Riemann hypothesis, Polignac's and Twin prime conjectures are Nonliving Things obeying Complex Elementary Fundamental Laws. People have often strived to obtain pivotal scientific answers on whether Living Things arise from Nonliving Things via the Evolution process [as per atheist belief] or Living Things arise from Nonliving Things via the Creation process [as per religious belief]. We speculatively hope and selfishly dream that the applied mathematics pathway resulting from solving Riemann hypothesis, Polignac's and Twin prime conjectures could at least one day lead to answering the

85

following question: Could the concocted expression "Living Things seem to exist at the edge of Chaos and Fractals" be mathematically equivalent to the following statement "Living Things must be made up of a combination of Completely Predictable entities, Incompletely Predictable entities, and Completely Unpredictable entities"?

Without going into finer details using Number theory, the irrationality measure (or irrationality exponent or approximation exponent or Liouville-Roth constant) of any real number is a measure of how "closely" it can be approximated by rationals. For a rational number, the irrationality measure is 1. The Thue-Siegel-Roth theorem states that for an algebraic irrational number, viz. real but not rational number, then the irrationality measure is 2. Transcendental irrational numbers have irrationality measure 2 or greater; for instance, the transcendental Euler's number e (= 2.718281828459...) has irrationality measure equal to 2. The [seemingly] simplistic-looking Liouville numbers is typified by Liouville's constant, sometimes also called Liouville's number, a

real number defined by $L \equiv \sum_{n=1}^{\infty} 10^{-n!}$ = 0.110001000000000000000001... with '!' denoting factorial. These numbers are irrational numbers of [the relatively more "complex"] transcendental types instead of [the relatively less "complex"] algebraic types; and their numerical make-up consist of just '0' and '1' digits. Despite this apparently simple-looking numerical make-up of Liouville numbers (as opposed to more complicated-looking numerical make-up of e), they are precisely those numbers [paradoxically] having infinite irrationality measure. For the above, we would assign all [Completely Predictable] rational numbers to obeying Simple Elementary Fundamental Laws, and all [Incompletely Predictable] irrational numbers to obeying Complex Elementary Fundamental Laws.

We now endeavor to compare, contrast and reconcile the two entities Living Things and Nonliving Things. Rigorous mathematical proofs must obviously be associated with 100% certainty. This can only apply to Simple and Complex Elementary

Fundamental Laws on Nonliving Things. Diverging onto proofs for Simple and Complex Emergent Fundamental Laws on Living Things, one observe that they can never be associated with perfect 100% certainty simply because we are dealing with "ALIVE" Living Things with dynamic spatial and temporal properties that could not be totally predictable. In this setting, the proofs for the Simple cases [e.g. physiologically modeling Cardiac Output (CO) equals to Heart Rate (HR) multiplied by Stroke Volume (SV) in the Cardiovascular System (CVS)] will comparatively be less challenging to derive than the Complex cases [e.g. physiologically modeling complex Human Brain functions using Neural Networks in the Central Nervous System (CNS)].

Note that the terms Elementary and Emergent are used here in the preceding and subsequent paragraphs to, respectively, denote Nonliving Things and Living Things. In real life situation for Living Things, there will always be the perpetual presence of infinitesimally tiny and unpredictable "Chaos and Fractals physiological variability", for instance, in the Simple Emergent Fundamental Law CO = HR X SR. This variability phenomenon will inevitably occur even in the most relaxed state of a person in deep sleep whereby dynamic processes such as intrinsic neuro-endocrine continuous signal input to the heart must occur on a permanent basis thus giving rise to this variability.

For the medically oriented readers, we finish off this topic by touching on Evidence based Medicine (EBM) and Evidence based Practice (EBP). Both could comply with either Simple or Complex Emergent Fundamental Laws on Living Things (namely, Human Beings in this scenario). EBM is typically depicted pictorially as a 'Pyramidal hierarchy of Literature Review' classifying available medical research materials into [the most powerful] Systematic Reviews down to [the least powerful] Expert Opinion.

Then EBP = Clinician Experience + Patient Expectation + Best Practice; with Best Practice being roughly equated with EBM. For doctors and medical researchers confronted daily with responsibly abiding to and improving up-to-date EBP and EBM, they must be familiar with most statistical tools employed in medical research. The classic example is research hypothesis expressed as a null hypothesis [the "devil's advocate" position] and alternative hypothesis. The level of statistical significance for hypothesis testing is often expressed as the so-called p-value. Whilst there is relatively little justification why a [cut-off] significance level of 0.05 is widely used in academic research [rather than 0.01 or 0.10]; we could be particularly more confident in our results by setting a more stringent level of (say) 0.01 [a 1% chance or less; 1 in 100 chance or less]. Despite this experimental/research tactic, we could strive to, but never, achieve perfect or 100% confidence in our results by setting ever more stringent levels.

We point out that there are overlapping pure and applied mathematics in our rigorous proof for Riemann hypothesis which was proposed more than 150 years in 1859. Conforming to logical arguments above, one could postulate that the lengthy delay in solving this hypothesis is simply because Riemann zeta function contains an infinite number of Incompletely Predictable intercepts demonstrating **Supraminimal Simplicity**, or alternatively stated, contains none of the Completely Predictable infinite intercepts demonstrating **Supramaximal Simplicity** [whereby Supramaximal Simplicity does allow multiple type solutions to prove a particular conjecture]. This will then require a proviso that there is only one [solitary] way using the "Complex Container Research Method" to solve this 'Incompletely Predictable problem' which belongs to the 'Special-Class-of-Mathematical-Problems with Solitary-Proof-Solution'.

Brief discussions on Statistics

I gratefully base the following brief discussions on statistics on materials present in HyperStat Online Statistics Textbook http://davidmlane.com/hyperstat/index.html, 1993 – 2013 by David M. Lane. Measurement is assignment of numbers to objects or events in a systematic fashion. Four levels of measurement scales are commonly distinguished: nominal, ordinal, interval, and ratio.

There is a relationship between level of measurement and appropriateness of various statistical procedures. For instance, it is foolish to compute mean of nominal measurements. Also, appropriateness of statistical analyses involving means for ordinal level data is controversial. One position is that data must be measured on an interval or a ratio scale for computation of means and other statistics to be valid. Thus, if data are measured on an ordinal scale, median but not mean can serve as measure of central tendency.

The arguments on both sides of this issue is examined in context of a hypothetical experiment designed to determine whether people prefer to work with color or with black and white computer displays. Twenty subjects viewed black and white displays and 20 subjects viewed color displays.

Displays were rated on a 7 point scale where a 1 was the lowest rating and a 7 was the highest rating. This rating scale is only an ordinal scale since there is no assurance that difference between a rating of 1 and a rating of 2 represents same degree of difference in preference as difference between a rating of 5 and a rating of 6.

The mean rating of color display was 5.5 and the mean rating of black and white display was 3.9. The first question experimenter would ask is how likely is it that this big a difference between means could have occurred just because of chance factors such as which subjects saw black and white display and which subjects saw color display. Standard methods of statistical inference can answer this question. Assume these methods led to conclusion that the difference was not due to chance but represented a "real" difference in means. Does the fact that rating scale was ordinal instead of interval have any implications for validity of the statistical conclusion that difference between means was not due to chance?

The answer is unequivocally 'no' with no room for argument here. What can be questioned is whether it is worth knowing that mean rating of color displays is higher than mean rating for black and white displays.

The argument that it is not worth knowing assumes that means of ordinal data are meaningless. Supporting notion that means of ordinal data are meaningless is the fact that examples can be made showing that a difference between means on an ordinal scale can be in opposite direction of what they would have been if "true" measurement scale had been used.

If means of ordinal data are meaningless, why should we care whether difference between two meaningless quantities (the two means) is due to chance or not. We answer this by challenging proposition that means of ordinal data are meaningless. Counter arguments or counter examples are often used to disprove a mathematical proposition, conjecture or hypothesis. There are two counter arguments to the example showing that using an ordinal scale can reverse the direction of difference between means.

The first is philosophical and challenges validity of the notion that there is some unseen "true" measurement scale that is only being approximated by rating scale. The second counter argument accepts the notion of an underlying scale but considers examples to be very contrived and unlikely to occur in real data. Measurement scales used in behavioral research are invariably somewhere between ordinal and interval scales. In preference experiment, it may not be the case that difference between ratings one and two is exactly the same as difference between five and six, but it is unlikely to be many times larger either. The scale is roughly interval and it is exceedingly unlikely that means on this scale would favor color displays while means on the "true" scale would favor black and white displays.

There are case examples where one can validly argue that use of an ordinal instead of a ratio scale seriously distorts conclusions. Consider an experiment designed to determine whether 5-year old children are more distractible than 10-year old children.

Measurement Scales

	No Distraction	Distraction
5-yr	6	3
10-yr	12	8

It appears as though 10-year olds are more distractible since distraction cost them 4 points (12 minus 8) but only cost 5-year olds 3 points (6 minus 3). Thus, it might be that a change from 3 to 6 represents a larger difference than a change from 8 to 12. We must consider the performance of 5-year olds dropped 50% (3/6 X 100%) from distraction but the performance of 10-year olds dropped only 33% (4/12 X 100%).

Which age group is "really" more distractible? Unfortunately, there is no right or wrong answer here. If proportional change is considered, then 5-year olds are more distractible; if amount of change is considered then 10-year olds are more distractible. We must keep in mind that statistical conclusions are not affected by choice of measurement scale even though important interpretation of these conclusions can be.

In above example, a statistical test could validly rule out chance as an explanation of finding that 10-year olds lost more points from distraction than did 5-year olds. However, statistical test will not reveal whether a greater drop necessarily means 10-year olds are more distractible. So, the conclusion that distraction costs 10-year olds more points than it costs 5-year olds is valid. Thus the interpretation depends on measurement issues.

Broadly speaking, statistical analyses provide conclusions about the numbers entered into them. Relating these conclusions to the substantive research issues depends on measurement operations.

Example for Measurement Scales – Assume there were a "true" measurement scale for job satisfaction and that it maps onto a 7-point rating scale as follows:

"True scale" 7-point scale

1 - 5	1
6 - 40	2
41 - 42	3
43 - 75	4
76 - 90	5
91 - 94	6
95 - 100	7

Then if someone's "true" job satisfaction were 55 he or she would have a rated score of 4. Now consider following two sets of job satisfaction scores:

	Group A		Group B	
	True Scale	Rating	True Scale	Rating
	6	2	5	1
	6	2	40	2
	43	4	74	4
	91	6	90	5
	95	7	100	7
Mean	48.2	4.2	61.8	3.8

On "true" scale mean for Group B is 61.8 which is much higher than mean for Group

A which is 48.2. However on the 7-point rating scale, mean for B is only 3.8 which is lower than mean for A of 4.2.

From my medical research conducted during my Anesthesia training viz. "Supramaximal elevation in B-type natriuretic peptide and its N-terminal fragment levels in anephric patients with heart failure: a case series" Journal of medical case reports, Primary author: (Ting J. Y., 2012) and Secondary author: (Pussell B. A., 2012); this study achieved the primary endpoint of demonstrating (sustained) supramaximal elevations of BNP and NT-proBNP in only one (obtainable) anephric patient inflicted with congestive heart failure (CHF) which suggested the need for dramatically higher BNP and NT-proBNP cut-off values for anephric patients in CHF with respective magnitudes of the order of 50-fold to 100-fold higher than the usual figures quoted to 'rule in' CHF.

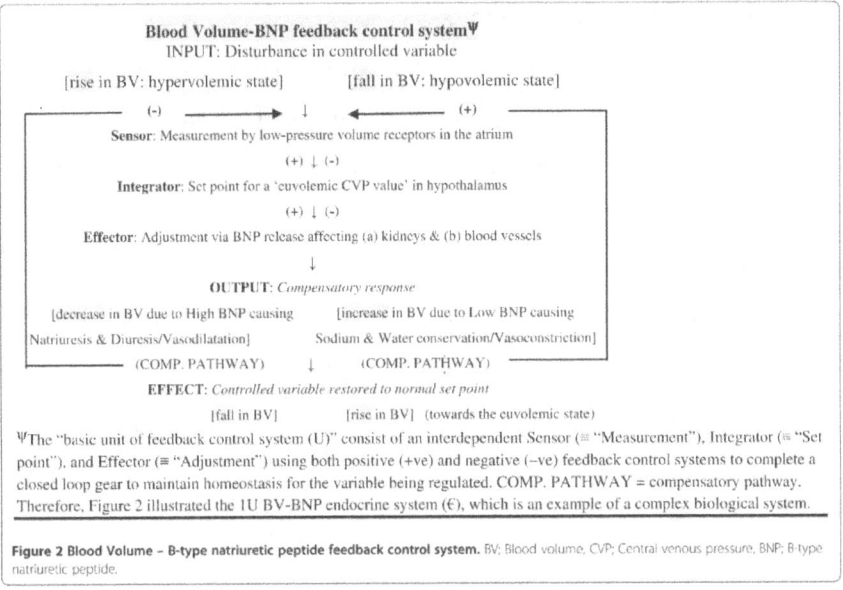

Figure 2 Blood Volume – B-type natriuretic peptide feedback control system. BV; Blood volume, CVP; Central venous pressure, BNP; B-type natriuretic peptide.

Introduction: CHF and renal failure (RF) act synergistically to increase the levels of B-type natriuretic peptide (BNP) and its co-secreted biologically inactive N-terminal fragment (NT-proBNP). These two cardiac neuro-hormones are mainly secreted from the ventricles and, to a lesser extent, the atria. They have an established role as useful diagnostic tests for CHF in both the pediatric and adult population, including RF patients.RF and CHF represent two merging pathologies with a varying spectrum for speed of onset and severity. The intersection of cardiac and renal insufficiency is referred to as cardiorenal syndrome (Type 1 to 5), which is CHF as a result of RF or vice versa. Cardiac dysfunction in end-stage kidney disease (ESKD) patients, whether acute or chronic, is often due to disorders of perfusion (ischemic heart disease) or to disorders of structure and function. The disorders of structure and function and CHF are often collectively termed 'uremic cardiomyopathy' which is commonly associated with left ventricle (LV) hypertrophy secondary to volume overload and hypertension (HT). Cardiac disease accounts for >50% of deaths in patients with End-Stage Kidney Disease (ESKD).

In theory, the truly anephric state represents a unique position for research purposes because any 'distorting interferences' from a failing kidney are eliminated from consideration. These interferences refer mainly to the kidney (an integral component of the feedback control system) acting as: (1) an &/or the end target-organ for the relevant hormones; and/or (2) an &/or the organ contributing to metabolic clearance rate (MCR) for the relevant hormones. This 'significant elimination' holds true for the first interference (with the associated loss of the kidney compensatory pathway) but less so for the second interference because: (i) anephric patients are dependent on, usually, intermittent hemodialysis (HD) to keep them alive and thus providing them with an artificial means to intermittently and variably clear relevant hormones (namely, BNP and NT-proBNP); and (ii) there may be non-renal pathways (via other organs or tissues) variably contributing to MCR for these relevant hormones as well.

Predominantly based on concepts behind the feedback control systems, we devise the interesting 'Blood Volume (BV) – BNP feedback control system' (Figure 2) [reproduced with permission], with further discourse given below to help provide plausible explanations for our findings.

Figure 2 which depicted a typical 1 U Blood Volume– BNP feedback control system, whereby the symbol 'U' stands for the 'basic unit of feedback control system'. The control of a physiological 'state' or 'variable' such as BV is via a complex web utilizing adaptive and integrative mechanisms. The immediate control of the total body water (TBW) endocrine system (€TBW) is predominantly mediated by the 'thirst-antidiuretic hormone (ADH) mechanism'. Thirst affects the input of water and ADH affects the output of water. The delayed control of TBW (a minor control) is mediated by the endocrine system (€s) such as the renin-angiotensin-aldosterone system (RAAS, which also mediates 'control of Blood Pressure (BP)' to some degree). The unit of this €TBW consists of: (i) sensors = osmoreceptors (monitor tonicity), low-pressure volume receptors (monitor BV), and high-pressure baroreceptors (monitor BP); (ii) integrator = hypothalamus; and (iii) effectors = thirst and ADH. Therefore the 'control of TBW' would also directly or indirectly lead to 'control of BP' (and vice versa) because both variables are controlled by RAAS to various degrees.

Similarly, the 'control of BV' variable (mediated by natriuretic peptides such as BNP in conjunction with RAAS) is affected to a varying degree by the 'control of TBW' and 'control of BP' variables. The '$\uparrow/\downarrow$ BV → $\uparrow/\downarrow$ myocardial stretch or tension → $\uparrow/\downarrow$ BNP release' homeostatic mechanism leads to the end-target organ effects of: (a) $\downarrow/\uparrow$ BV (via (natriuresis and diuresis) / (sodium and water conservation) by kidneys) and (b) $\downarrow/\uparrow$ peripheral vascular resistance (via (vasodilatation) / (vasoconstriction) on blood vessels). This signifies that our proposed homeostasis (with integrator = hypothalamus likely connected via autonomic neural pathway) acting through its compensatory pathways is invoked to help restore the disturbance in BV ('hypervolemic' / 'hypovolemic') to its 'euvolemic' set point resulting in improved diastolic relaxation (lusitropy) and decreased myocardial fibrosis. Measurements on BV status are largely carried out by the low-pressure volume receptors in the atrium (which essentially equates to central venous pressure clinical measurements). Likewise, detailed analysis for a 2 U hypothalamic–pituitary axis for prolactin hormone or a 3 U hypothalamic–pituitary–thyroid axis for thyroid hormones could also be carried out.

For a given €, the magnitude of its size and complexity would increase exponentially if a linear increase in the number of U were to occur due to the associated power-law increase in the number of 'controlled variables' along with their 'mini-components' (input, output, effect and compensatory pathway). The total number of possibilities arising from n mini-components when considering the '($\uparrow$ or +)' or '($\downarrow$ or –)' state (i.e. r = 2) for each mini-component is given by the permutations with repetition formula: n^r = n^2 from combinatorics. Let the symbol Σ denote 'the sum of'; and x_i and y_i denote 'n individual factors or causes of endocrine disorder' for i = 1, 2, 3,..., n that tend to have elevating or lowering properties, respectively, on the relevant hormone.

The overall magnitudes of rises or falls in the particular hormonal output (O) (which is a controlled variable), such as BNP, NT-proBNP and prolactin hormonal concentrations, are governed by the net difference between the resultant effect from Σ (effects from x_i that tend to increase O) and Σ (effects from y_i that tend to decrease O). This overall resultant effect stemming from the absolute difference between the n value for x_i (n_x) and the n value for y_i
(n_y), namely $|n_x - n_y|$, that tends to increase or decrease O respectively would be some nonlinear function of this absolute difference of a synergistic nature. Then this overall resultant effect will be of ever greater cumulative rises or falls (of an exponential nature) in O when $|n_x - n_y|$ is numerically >1 and constitutes an ever larger integer number.

The '$\uparrow$/$\downarrow$ BV causing $\uparrow$/$\downarrow$ myocardial stretch or tension, resulting in $\uparrow$/$\downarrow$ BNP release' is the main mechanism for BNP (and NT-proBNP) pulsatile co-secretion. Other mechanisms such as heart muscle cell damage from myocardial infarct will also lead to BNP and NT-proBNP release.

There are two major cardiac and non-cardiac causes of BNP and NT-proBNP elevations as follows:

First, moderate increases in BNP (100–500ng/L) or NT-proBNP (250–1000ng/L): ventricular dysfunction, ischemic heart disease, pulmonary HT, acute pulmonary embolism, cor pulmonale, septic shock, renal insufficiency, liver cirrhosis, subarachnoid hemorrhage and hyperthyroidism.

Second, severe increases in BNP (>500ng/L) or NT-proBNP (>1000ng/L): decompensated heart failure (HF), pulmonary Hypertension (HT), acute pulmonary embolism and septic shock.

In addition to glomerular filtration, BNP is eliminated from plasma mainly through natriuretic peptide receptors and degraded by neutral endopeptidases. By contrast, it is possible that NTproBNP is largely eliminated by glomerular filtration. Levels of both BNP and NT-proBNP are: elevated with ageing, higher in women than in men, higher in RF and CHF of greater severity, and higher in LV systolic HF than LV diastolic HF. Stage of HF (early versus late) and genetic polymorphisms may result in inter-individual variation of BNP and NT-proBNP. Obesity with and without CHF is associated with lower levels of both molecules; obesity with and without CHF is presumably attributed to non-hemodynamic factors such as BMI-related defect in natriuretic peptide secretion (from either ↓myocardial hormone release or ↓synthesis), and ↑BNP metabolism in adipose tissue either via peptide degradation or regulation of clearance receptors.

The end target-organs for BNP are the kidneys and blood vessels. These are associated with the 'kidney compensatory pathway' and 'blood vessel compensatory pathway' respectively. The SIA state corresponds to the loss of the kidney as (a) an end target-organ and (b) a compensatory pathway, although the contribution of the collective blood vessels as an end target-organ and compensatory pathway is still intact. The primary endpoint of our study was to demonstrate the supramaximal elevation of BNP and NT-proBNP in Patient 1. Computing from Figure 2 (together with the 'major cardiac and non-cardiac causes of BNP and NT-proBNP elevations'), this can be seen to be due to multiple x_i (with no identifiable y_i); namely: (i) CHF itself, (ii) decreased MCR for BNP and NT-proBNP, and (iii) total loss of kidney tissue acting as an end-target organ (but intact collective blood vessels acting as an end-target organ) with total disruption of 'kidney compensatory pathway' loop.

The secondary endpoint of our study was to demonstrate the supramaximal elevation of prolactin in Patient 1 as suggested by the persistently high prolactin values obtained between Event X (development of acute CHF) and Event Y (death of the patient). This was due to multiple x_i (with no identifiable y_i); namely, the systemic disorders of: (i) chronic RF, (ii) emotional stress, (iii) epileptic seizures, (iv) pharmacologic factors (anti-HT dopamine synthesis inhibitors methyldopa), and (v) decreased MCR for prolactin. Both the mildly elevated prolactin levels in Patients 2 and 3 mainly reflect the decreased MCR of prolactin due to CKD Stage 3 and anephric status (needing intermittent HD) for each respective patient.

The $n_x = 3$ in Patient 1 for Σ (effects from x_i that tend to increase the 'outputs' of BNP and NTproBNP) with resultant massive and persistent elevation of these two natriuretic peptides. Applying the $n_x = 3$ minus $n_y = 0$ calculation giving a 'relatively large' $|n_x - n_y|$ value of 3 predicts the overall magnitude of rises to be consistent with the supramaximal elevation of these hormones as seen in our study. Similar calculation of a 'relatively large' $|n_x - n_y|$ value of 5 for prolactin x_i and y_i in Patient 1 also showed that they act in concert to greatly increase and maintain the high prolactin 'output' in a synergistic manner to explain its supramaximal elevation.

'Functional' anephric states should occur in adult ESKD patients when their in-situ remnant kidney tissues have totally lost all their functions or have atrophied completely. One could extrapolate that these patients should behave physiologically in a similar manner to surgically-induced anephric (SIA) patients. A corollary to this argument would result in the hypothesis that when ESKD patients develop CHF with supramaximal elevations of BNP and NT-proBNP, they are likely to be functionally anephric. The full significances of this hypothesis in adults are yet to be fully realized.

Supramaximal hormonal elevations when observed in neonates, infants and children will undoubtedly be due to mechanisms similar to that of their adult counterpart; and with the full impact of this hypothesis lying in uncharted territories. These are exciting areas for future medical research. Let us mathematically analyze the following statement in a logical manner: The defined parameters n_x, n_y, and $|n_x - n_y|$ are applicable to both 'anatomical' and 'functional' anephric patients. Because one can safely assume that all supporting criteria for the statement to hold true are present in both sets of patients, then this 'common denominator' statement per se can provide intuitive non-contradictory explanations for the supramaximal elevation phenomenon in all anephric patients inflicted with CHF. This 'common denominator' statement thus lends support to our proposed hypothesis that 'anatomical' and 'functional' anephric patients inflicted with CHF should have similar natriuretic response behavior.

Footnote on 'An infant in temporary anephric and congestive heart failure state manifesting supramaximal elevations of natriuretic peptides'

In October 2010, we encountered the case of a 5-month-old male baby (weight 7kg) with out-of-hospital cardiac arrest (due to commotio cordis) requiring 30 minutes of cardiopulmonary resuscitation before return of spontaneous circulation. He developed multiple organ dysfunction syndrome (MODS) requiring full Intensive Care Unit (ICU) supportive care. The ICU supportive care included therapeutic hypothermia between 33°C and 34°C for the first 48 hours, full invasive ventilation for 12 days for acute CHF with fractional shortening (FS) 31% on echocardiogram (normal >30%) while on multiple inotropic and vasopressor agents, and Continuous Veno-Venous Hemodiafiltration (CVVHDF) for 5 days from Day 3 to 7 for (anuric) acute kidney injury with peak creatinine 93μmol/L (20–50) on Day 8: this probably equates to the baby being a temporary 'functional' anephric patient. Blood tests on Day 7 showed: Hb 88g/L, creatinine 50μmol/L, and supramaximal elevation of NT-proBNP at 173,392ng/L (20,460pmol/L). By Day 10, the patient had not required dialysis for 3 days with blood tests showing: Hb 84g/L, creatinine mildly elevated at 58μmol/L (with good urine output), and lesser magnitude of supramaximal elevation of NT-proBNP at 98,476ng/L (11,620pmol/L). He continued to steadily improve before being extubated onto continuous positive airway pressure on Day 12 with eventual discharge from the pediatric ICU on Day 27. Follow-up of the patient at 12 months post-cardiac arrest revealed remarkable recovery from his MODS with possibly very mild and subtle residual cognitive dysfunction from the hypoxic ischemic encephalopathy, normal renal function (creatinine 23μmol/L at 8 weeks post-arrest), and normal cardiac function (Fractional Shortening 42% on echocardiogram at 7 months post-arrest).

General Chemistry: Stoichiometry

An important field of chemistry is stoichiometry, which is the quantitative relationship between chemical substances in a reaction. The weight of a molecule is the sum of weights of atoms of which it is made. The unit of weight is **dalton (Da)**, one-twelfth the weight of an atom of ^{12}C, and 1000 Da = 1 kilodalton (kDa). The molecular weight of a substance is the ratio of mass of one molecule of substance to mass of one-twelfth the mass of an atom of ^{12}C – being a ratio, it is dimensionless. A **mole** is the gram-molecular weight of a substance, that is, the quantity of a substance whose weight in grams in equal to molecular weight of the substance. There are approximately 6 x 10^{23} molecules in a mole. (This is Avogadro's Number with best experimental value of 6.02214199 x 10^{23} atoms per mol.)

One very important property of solutions that must be addressed is concentration. Concentration generally refers to the amount of solute contained in a certain amount of solution. To deal with concentration you must keep in mind the distinctions between solute, solvent and solution. There are five units of concentration that are particularly useful to chemists. The first three: **molarity (with its closely related osmolarity)**, **molality (with its closely related osmolality)** and **normality** are dependant upon the mole unit. The last two: **percent by volume** and **percent by weight** have nothing to do with mole, only weight or volume of the solute or substance to be diluted, versus the weight or volume of the solvent or substance in which the solute is diluted. (Note that Percentages can also be determined for solids within solids.)

Molarity is the number of moles of solute dissolved in one liter of solution. The units are moles per liter of solution (**M**). A **molality** is the number of moles of solute dissolved in one kilogram of solvent. The units are moles per kilogram of solvent (**m**). Note that to prepare a given molality solution, the solvent must be weighed unless it is water. (One liter of water has a specific gravity of 1.0 and weighs one kilogram; so one can simply measure out one liter of water and add the solute to it. Most other solvents have a specific gravity greater than or less than one; thus one must weigh the solvent.)

The diffusion of solvent molecules into a region in which there is a higher concentration of a solute to which the membrane is impermeable is called osmosis. The pressure necessary to prevent solvent migration is the osmotic pressure of the solution. Like vapor pressure lowering, freezing-point depression, and boiling-point elevation, osmotic pressure depends upon the number rather than the type of particles in a solution – in other words, it is a fundamental colligative property of solutions. In an ideal solution, osmotic pressure (P) is related to (the absolute) temperature (T) and volume (V) in the same as the pressure of a gas: $P = nRT/V$, with n the number of particles and R the gas constant. The concentration of osmotically active particles is expressed in **osmoles**, with one osmole (osm) equals the gram-molecular weight of a substance divided by the number of freely moving particles that each molecule liberates in solution. (One milliosmole (mosm) is $1/1000$ of one osm.) Our body fluids is not an ideal solution and although the dissociation of the solute made of an ionizing compound (such as strong electrolytes) is complete, the number of particles free to exert an osmotic effect is reduced owing to interactions between the ions – thus, it is the effective concentration (*activity*) in the body fluids rather than the number of equivalents of an electrolyte in solution that determines its osmotic effect, and this deviation from an ideal solution is greater with the more concentrated the soultion. **Osmolarity** defined as the number of osmoles per liter of solution (osm/L). If the solute is a nonionizing compound, the osmolarity is the same as molarity – otherwise (as per reasoning above) in the solute made of an ionizing compound, this is not true with the osmolarity greater than the molarity. The term tonicity is used to describe osmolality of a solution relative to plasma such that solutions that have same osmolality as plasma are said to be isotonic; those with greater osmolality are hypertonic; and those with lesser osmolality are hypotonic. With respect to argument for whether a solute is made up of an ionizing or a nonionizing compound, a similar line of thinking can be applied for **osmolality**, which is defined as number of osmoles per kilogram of solvent (osm/kg) – thus, osmolality is only the same as molality in the non-ionizing compound case and they are different in the ionizing compound case. *Note that osmolarity is affected by the volume of various solutes in the solution and the temperature, while osmolality is not.*

The concentration of a solution can be stated by the amount of solute in equivalents rather than moles. This is called **normality (N)**, which is the number of equivalents of solute per liter of solution (eq/L). Electrical equivalence is not necessarily the same as chemical equivalence. One (electrical) equivalent (eq) is 1 mol of an ionized substance divided by its valence. One (chemical) gram equivalent is the weight of a substance that is chemically equivalent to 8.000 g of oxygen. Therefore, the normality (**N**) of a solution can also be expressed as the number of gram equivalents in 1 liter (g/L). For example, one mole of HCl dissociates into 1 eq of H^+ and 1 eq of Cl^-. One equivalent of $H^+ = 1$ g/L $= 1$ g and 1 eq of $Cl^- = 35.5$ g/l $= 35.5$ g. Thus, a 1 M (mol/L) solution of hydrochloric acid contains 1 mol/L x 2 eq/mol $= 2$ N (eq/L) $= 1 + 35.5$ g/l $= 36.5$ g/L. There is a close relationship between normality and molarity. Normality can only be calculated when we deal with reactions, because normality is a function of equivalents. Normality = molarity x n, where n (eq/mol) = the number of protons (or hydrogen ions) exchanged in a reaction. In other words, the normality of a solution is simply a multiple of the molarity of the solution. Generally, the normality of a solution is just one, two or three times the molarity. In rare cases it can be four, five, six or even seven times as much. Applying the same above line of thinking all over again, we can easily conclude that osmolarity only approximates normality in the case of the ideal solution.

Percentages are easy to calculate because they do not require information about the chemical nature of the substance. Percentages can be determined as percent by weight or percent by volume. Percentages are used more in the technological fields of chemistry (such as environmental technologies) than they are in pure chemistry.

Percent by weight: To make up a solution based on percentage by weight, one would simply determine what percentage was desired (for example, a 20% by weight aqueous solution of sodium chloride) and the total quantity to be prepared. If the total quantity needed is 1 kg, then it would simply be a matter of calculating 20% of 1 kg which, of course is: 0.20 NaCl * 1000 g/kg = 200 g NaCl/kg. In order to bring the total quantity to 1 kg, it would be necessary to add 800g water.

Percent by volume: Solutions based on percent by volume are calculated the same as for percent by weight, except that calculations are based on volume. Thus one would simply determine what percentage was desired (for example, a 20% by volume aqueous solution of sodium chloride) and the total quantity to be prepared. If the total quantity needed is 1 liter, then it would simply be a matter of calculating 20% of 1 liter which, of course is: 0.20 NaCl * 1000 ml/l = 200 ml NaCl/l.

The Stewart hypothesis of acid-base balance focus on definitions of acid and base. In order to understand the differences between the classic ("Siggaard-Anderson") approach to acid-base analysis and the new paradigm Stewart approach, it's helpful to define a few points from which to work. Information content in this section have gratefully been based upon materials by Dr. Pete Watkinson from relevant AnaesthesiaUK website on acid-base balance [https://www.frca.co.uk/article.aspx?articleid=100924].

The first important point is how to define an acid. This is key because from the definition comes the term "acidotic". Most people would think of a patient as being acidotic when the pH is below 7.35. As pH is purely a function of the theoretical hydrogen ion concentration of a solution $-log_{10}[H+]$, an acid is being defined as a substance that can donate a proton. This simple definition is in line with modern theory.

There are, however, various other ways of defining an acid. An acid can also be defined as any substance that produces an increased concentration of hydrogen ions when dissolved in water (the 'Arrhenius definition'). The Stewart approach approximates this definition. Not surprisingly, attachment of different meanings to the same terminology has led to considerable confusion throughout the history of acid base physiology.

The second important point to note is that both the Siggard-Anderson and Stewart approaches to acid base analysis are mathematical models. Both approaches do not explain the mechanism by which a person has become biochemically acidotic/alkalotic, nor do they claim to do this, but both models merely attempt to illustrate the area the acid-base disturbance lies.

We now focus on the Siggaard-Anderson approach whereby in the Siggaard-Anderson model, an acid-base disturbance is looked at using a combination of 3 factors:
    # The Henderson-Hasselbach equation
    # The base excess

The anion gap.

The Henderson-Hasselbach equation comes from the dissociation equation for carbonic acid, and its use is based on the premise that in normal plasma, bicarbonate is the most important buffer.

From this equation, the only two factors affecting pH are:
bicarbonate concentration
pCO_2

The Henderson-Hasselbach equation provides an approximate relationship between respiratory variable (pCO_2), metabolic variable [HCO_3-], and resultant pH. A flaw with this approach is that other important buffers exist and play an important role in acid base physiology (e.g. hemoglobin and albumin). HCO_3- and pCO_2 are not therefore independent. In the simplified dissociation equation below, a rise in pCO_2 will cause dissociation to shift to the right as a result of the law of mass action:

$$CO_2 + H_2O \leftrightarrow H+ + HCO_3-$$

Protons will be buffered by hemoglobin and albumin, and the bicarbonate levels will rise. So a rise in pCO_2 has resulted in a rise in [HCO_3-]. The rise in [HCO_3-] could easily be mistaken as a metabolic alkalosis, when in fact the true cause was a respiratory acidosis. The base excess concept was evolved to address this problem. It is a method of measuring the metabolic component. The base excess concept works by resetting the sample to a normal pCO_2 (5.33kPa) by equilibration, and then titrating it to pH 7.4 using molar acid (now calculated from normograms). The number of mmol /L required equals the base excess, and is therefore a measure of how acidotic or alkalotic the sample is without any contribution of carbon dioxide.

Finally, calculation of the anion gap (see Equation box below) allows classification of a metabolic acidosis into those with a normal or increased anion gap. The anion gap is a measure of the concentration of unmeasured anions (e.g. plasma proteins) and is based on the theory of electrical neutrality (the sum of the positive ions must equal the sum of the negative ions). An increased anion gap suggests the presence of unmeasured organic acid, whereas a normal anion gap implies that the decrease in bicarbonate has been counteracted by an increase in chloride concentration (see following Table).

Table: Causes of metabolic acidosis - Siggard-Anderson approach

Increased anion gap (usually decreased [Cl-]	Normal anion gap (usually increased [Cl-])
Ketoacidosis	Diarrhoea
Alcoholic	Parenteral nutrition
Diabetic	Carbonic anhydrase inhibitors
Starvation	Dilutional acidosis
Hyperosmolar nonketotic coma	Ingestion of HCl or other acid
Lactic acidosis	Renal tubular acidosis
Uraemic acidosis	Ileostomy
Methanol	Ureterosigmoidostomy
Ethylene glycol	Pancreatic fistula
Salicylate	
Paraldehyde	

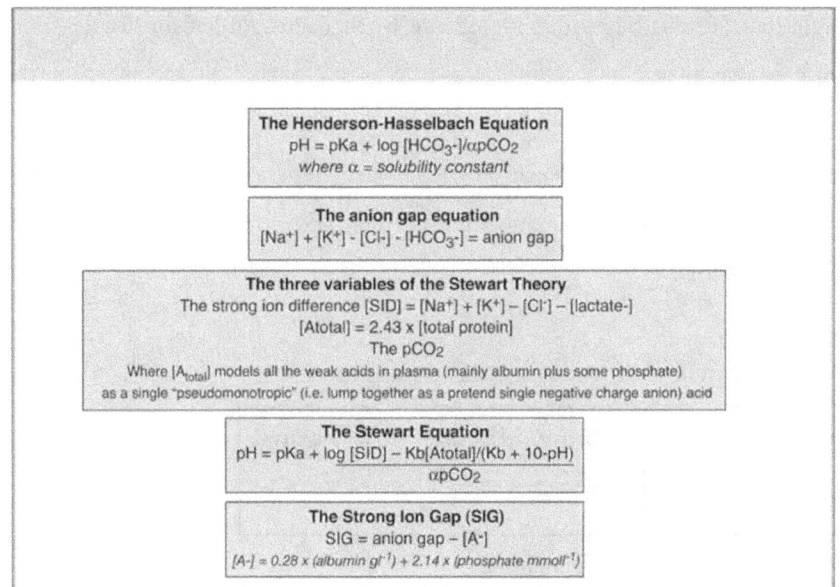

Equati

on Box for Stewart approach.

Stewart took the same system, but looked at from a slightly different angle. He concluded that one might model acid-base disturbances, based on three conceptual contributors described in Equation box above:

1. The Strong Ion difference (SID)
2. Weak Acids in Plasma (A total)
3. PCO2

The law of electrical neutrality means that:

$[Na+] + [K+] + [H+] = [Cl-] + [lactate-] + [HCO3-] + [A-] + [CO32-]$

Ignoring the minimal contribution of $[H+]$, $[HCO3-]$ and $[CO32-]$, and substituting the strong ion difference shows:

$[SID] = [HCO3-] + [A-]$

Stewart puts the three variables together in the Stewart Equation described in the equation box. It is interesting to note that if you ignore the contribution of albumin in this equation, it simplifies to the Henderson-Hasselbalch equation. Thus, albumin is the major variable that Stewart has added in, left out by Siggaard-Andersson for reasons of simplicity.

From this comes a practical utilization of Stewart theory in that it can be used to define the Strong Ion Gap (see equation box) which allows metabolic acidosis to be classified in Table below.

Table: Causes of metabolic acidosis – Stewart approach

Low SID and high SIG	Low SID and low/normal SIG
Ketoacidosis	Diarrhoea
Alcoholic	Parenteral nutrition
Diabetic	Carbonic anhydrase inhibitors
Starvation	Saline/dilutional
Lactic acidosis	Renal tubular acidosis
Methanol	Ileostomy
Ethylene glycol	Ureterosigmoidostomy
Salicylate	Pancreatic fistula

When comparing the two tables above using Sigaard Anderson approach and Stewart approach, they illustrate that the two major classifications are broadly similar.

Let us focus on a real life example of a metabolic acidosis. The Sigaard Anderson and Stewart models of acid base disturbance can be illustrated by the following case, commonly encountered in ICU. The table below shows the blood gas analysis of an elderly patient admitted to ICU from operating theatre after a prolonged laparotomy for which he was given 6 litres of saline during the operation:

Table: Commonly encountered post-operative Arterial Blood Gas

ABG	Admission to ICU	Discharge from ICU
Saturation%	95	97
pH	7.28	7.35
pO2	9.7	11.2
pCO2	5.7	5.2
Base excess	-6	-1
Sodium mmol/L	145	138
Potassium mmol/L	4.3	3.9
HCO3- mmol/L	19	24
Chloride mmol/L	115	105
Lactate mmol/L	1.3	1.2
Glucose mmol/L	7.1	6.4

In Siggaard-Anderson's model, the patient has a metabolic acidosis with a normal anion gap. Given the history, a diagnosis of "dilutional" due to saline is the likely explanation.

In the Stewart model, calculation of the SID shows it to be low (normal ~40mmol/L), and the SIG (normally ~0) is approximately normal, (assuming an albumin of 45g/L and a phosphate of 1 mmol/L):

[A-] = 0.28 X (albumin) + 2.14 X (phosphate) = 12.6 + 2.14 = 14.74

[SID] = [HCO3-] + [A-] = 19 + 14.74 = 33.74

Anion gap = [Na+] + [K+] − [Cl-] − [HCO3-] = 145 + 4.3 − 115 − 19 = 15.3

SIG = AG − [A-] = 15.3 − 14.74

Again, given the history, we draw the same conclusion. Neither of these models provide the mechanism by which saline has caused a significant acidosis. So in this case, the patient is in the group with either a decreased strong ion difference, or normal anion gap, depending on whose model we are using. In either case the patient is hyperchloraemic, as serum chloride falls outside the normal range (115 mmol/L). Neither model explains why or even if hyperchloridaemia is causatory in acidosis.

We now focus on the cause of hyperchloraemic acidosis which is thought to be due to a combination of factors. Firstly, normal saline (pH 5-6) possesses little buffering capacity and is being used to replace blood (pH of 7.4) characterized by extensive buffering capacity. The acidic pH of normal saline is due to a combination of dissolved CO_2 and an effect known as Grotthus mechanism (also known as proton jumping) whereby this is the process by which an 'excess' proton or proton defect diffuses through the hydrogen bond network of water molecules or other hydrogen-bonded liquids through the formation and concomitant cleavage of covalent bonds involving neighboring molecules. Secondly, volume expansion causes plasma bicarbonate dilution, and renal bicarbonate wasting.

In summary:

The two main approaches to acid base analysis are the Siggard-Anderson approach (the 'classic') and the Stewart hypothesis.

The three main determinants of acid base status according to Stewart hypothesis are the Strong Ion Difference (SID), bicarbonate (HCO3-) and albumin (A-).

Albumin is the major additional variable of acid base analysis that Stewart has considered.

Stewart's consideration of plasma protein levels acid base analysis is important for ICU patients as albumin levels are often disturbed. Avoidance of such large quantities of saline-containing fluids may help prevent the incidence of hyperchloraemic acidosis on the ICU patients.

The above discussion on 'modern' Stewart hypothesis during my Intensive Care Medicine and Anesthesia training from 2009 to 2013 was what I predominantly learn during that period. Acid-base analysis on human beings (Living Things) represent problems that involve Simple Emergent Fundamental Laws.

10 Mathematical Impossibility of E-Commerce Apocalypse (plus Pathology tests)

We have already mentioned prime counting function in substantial details in previous chapters. In relation to this function, other functions more convenient to work with can also be utilized and they open up a whole new world of marvelous mathematical relationships. An example is Riemann prime counting function (aka prime power counting function), commonly denoted by $J(x)$. This non-infinite series function has jumps of $1/n$ for prime powers p^n, and with it taking a value halfway between the two sides at discontinuities. Amazingly, the prime counting function $\pi(x)$ is related to by $J(x)$ by the Mobius transform. More amazingly still, $J(x)$ is related to Riemann zeta function through the Mellin transform (which is an integral transform).

The word "Apocalypse" comes from Greek, and actually means "to uncover, to reveal". But in modern times it has come to mean the final destruction that is revealed in the Holy Bible from the Book of Revelation (the last book of the New Testament). In the early vision of the Apocalypse, the final blow is a massive earthquake. "Then the spirits brought the kings together in the place that in Hebrew is called Armageddon" (Revelation 16:16 from the Good News Bible, 1988 Australian edition). In biblical terms, Armageddon is the scene of the last battle to be fought between the forces of good and evil, prophesied to happen at the end of time. But Armageddon is also a real place. It is the Greek name for an ancient city in Israel, Megiddo. In Hebrew, "Mount Megiddo" is "Harmegiddo", and "Armageddon" is simply the Greek transliteration of that name. There are thought to be five common scenarios on how our world could end:-

(1) Nuclear holocaust. (2) Doomsday virus. (3) Killer Asteroid. (4) Threat of global warming. (5) Doomsday volcano. Should we add "The Great Earthquake" as the sixth Doomsday machine? Perhaps we should be more optimistic and less pessimistic.

Can we find a book with hidden codes that is supposed to reveal prophesies and predictions for the future? If so, is such a profound allegation scientifically sound? Michael Drosnin in his book 'The Bible Code' (1997, Publisher: Weidenfeld & Nicholson, London) is claiming that decoding the Bible [more specifically, the first five books of the Hebrew bible] allegedly leads to the discovery of prophecies and profound truths of a secular nature, not all of which are related to the Jews. He proposes that the Bible is the only text in which these encoded phrases are found in a statistically significant pattern, and that the chance of this being a random phenomenon is unlikely. Using the equidistant letter sequences (ELS) method, he claims that the assassinations of Yitzhak Rabin, Anwar Sadat and the Kennedy brothers were foretold in the Bible.

The Bible (or Torah) Code is a code alleged to have been intentionally embedded in the Bible. The code is revealed by searching for equidistant letter sequences (ELS). Doron Witztum, Eliyahu Rips and Yoav Rosenberg published their findings in the journal Statistical Science (1994, Vol. 9, No. 3, 429-438) under the title of "Equidistant Letter Sequences in the Book of Genesis." When the authors used a randomization test to see how rarely the patterns they found might arise by chance alone they obtained a highly significant result, with the probability $p = 0.000016$. That is, the probability of getting the results they did was 16 out of one million or 1 out of 62,500. The authors commented that the randomization analysis shows that the effect is significant at the level of 0.00002 and the proximity of ELS's with related meanings in the Book of Genesis is not due to chance. The mathematics was felt "to be solid" and "the numbers held up".

Here is a simplified version on how the Bible Code is supposed to work. The idea is to start at a letter 'x' (which we can vary) and then to pick every nth letter while abiding by the rule (where n is also a number which we can vary). A computer program might read as follows: The starting point is letter 'x'. Input 'x' as the first letter, then 'x plus n' as the second letter, then 'x plus 2n' as the third letter, then 'x plus 3n' as the fourth letter, and so on. Using different values for "N" and "D", one can generate many strings of letters. Imagine wrapping the string of letters around a cylinder in such a way that all the letters can be displayed. Flatten the cylinder to reveal several rows with columns of equal length, except perhaps the last column which might be shorter than all the rest. Now search for meaningful names in proximity to dates. Search horizontally, vertically, diagonally, any which way. A group of Israeli mathematicians did just this and claimed that when they searched for names in close proximity to birth or death dates (as published in the Encyclopedia of Great Men in Israel) they found many matches. Analyze the eventual output result as a long string of letters and see if meaningful words or names might pop out. A group of Israeli mathematicians did just this and claimed that when they searched for names in close proximity to birth or death dates (as published in the Encyclopedia of Great Men in Israel) they found many matches. However, the consensus is that the Bible Code is basically a method of employing some hundreds, if not thousands, of millions of ways of "fishing out words at random, by starting anywhere you like and taking any steps you like".

With such a profound allegation made by Michael Drosnin, there are bound to be many critics. Therefore, in response to his critics, the following challenge was made by Drosnin: "When my critics find a message about the assassination of a prime minister encrypted in Moby Dick, I'll believe them." (Newsweek, 9 June 1997)

The commonly used public key encryption system RSA (Rivest-Shamir-Adleman) was first described in 1977 by Ron Rivest, Adi Shamir and Leonard Adleman from Massachusetts Institute of Technology. It gets its security from the difficulty of factoring large integers that are the product of two large prime numbers. Internet transactions in e-commerce depends on the integrity of humongous [non-prime] numbers to be anonymously constituted from its basic prime numbers such as that utilized by RSA. It is often thought that breaching this integrity by being able to easily identify prime numbers constituents of relevant humongous numbers after successfully solving Riemann hypothesis would have massive implication in that it will now brought the whole of e-commerce to its knees overnight.

The truthfulness of the preceding narrative paragraph can now be beautifully refuted in a mathematical manner. Solving Riemann hypothesis, Polignac's and Twin prime conjectures is simply irrelevant because the CIS of nontrivial zeros and prime numbers must be treated as Incompletely Predictable entities abiding by Complex Elementary Fundamental Laws that are "Incompletely Predictable Laws" [and not Simple Elementary Fundamental Laws that are "Completely Predictable Laws"]. Thus, in principle, we have dispelled the doom-and-gloom prophecy that financial disaster might follow when successful proof of Riemann hypothesis occur.

However, in practice, there may be a twist to this sentiment. Building ever more powerful supercomputers, which are classical computers based on classical physics, could more easily crack crptic codes but this issue can progressively be negated by employing ever larger prime numbers in cryptic codes. The world's smallest transitor made from a single atom was created in 2012. This should hypothetically assist in the future building of the most powerful supercomputers.

But the infinitely more powerful quantum computer, based on quantum mechanics phenomena such as superposition and entanglement, could solve problems in minutes which would otherwise take thousands of years due to the theoretical ability of quantum computers to do a huge range of calculations simultaneously rather than sequentially as in classical computers. This will revolutionize research into areas like artificial intelligence, self-driving cars and drug design in a positive manner but will likely impact the desired role of many cryptic codes in a negative manner by easily cracking them ("cryptocalypse").

Quantum computers could easily crack many crytic codes such as that used by RSA in polynomial time by using Shor's algorithm to find the prime number factors of large integers. To circumvent this problem, the use of alternative crytic codes such as the lattice-based cryptosystems which are known not to be broken by quantum computers will be desirable. Finally, 'quantum cryptography' (as opposed to 'classical cryptography') could potentially fulfill some of the functions of public key cryptography.

Pathology Tests

In this section, I will concentrate mainly on the more novel aspects associated with pathology tests that are usually not available in standard textbooks or, for that matter, have never been formulated before. In order to understand them, we will have to touch base with some basic familiar concepts that are usually known to most people.

Guidelines are often drawn up to help clinician manage a clinical condition. For instance, in the "Management of unstable angina guidelines", Medical Journal of Australia, Vol 173, S65-S88, 16 October 2000; the evidence in the guidelines is graded according to the level-of-evidence classifications endorsed by the National Health and Medical Research Council (NHMRC) in 1995. These are:

E1 Level I: Evidence obtained from a systematic review of all relevant randomized controlled trials.

E2 Level II: Evidence obtained from at least one properly designed randomized controlled trial.

E3 Level III: Evidence obtained from all well-designed controlled trials without randomization, well-designed cohort or case-control analytic studies, preferably from more than one center or research group, or from multiple time series with or without the intervention. Dramatic results in uncontrolled experiments (such as the results of the introduction of penicillin treatment in the 1940s) could also be regarded as this type of evidence.

E4 Level IV: Opinions of respected authorities, based on clinical experience, descriptive studies, or reports of expert committees.

Working clinicians often attempt to keep abreast of development by selectively reading those journal articles (among the exponentially expanding literature) that are relevant, valid and applicable. Research articles can be on a diagnostic test, the clinical course & prognosis of a disorder, the etiology or causation, and the usefulness of a therapy. The title, authors, summary and site of a research article will often quickly decide whether the article is worth-while reading. For instance, in targeting an article on a diagnostic test, once we decided that the article passes these four brief tests, only then do we proceed to read the "Patients and Methods" section.

Then in so doing, one of the most important elements of the proper clinical evaluation of a diagnostic test is "Was there an independent, 'blind' comparison with a 'gold standard' of diagnosis?" In answering this question, we will have come across important parameters (see below for definitions) from "stable properties" and from "frequency-dependent properties".

Gold standard:

		Positive	Negative
Test	Positive	**True +ve [a]**	**False +ve [b]**
Result:	Negative	**False –ve [c]**	**True –ve [d]**

Stable properties:
  Sensitivity (Sen) = a/(a+c)
  Specificity (Spec) = d/(b+d)

Frequency-dependent properties:

Positive predictive value (+ve Pred value) = a/(a+b)

Negative predictive value (-ve Pred value) = d/(c+d)

Accuracy (Accu) = (a+d)/(a+b+c+d)

Prevalence (Prev) = (a+c)/(a+b+c+d)

Using Bayes' theorem, +ve Pred values can also be calculated as

[(Prev) (Sen)]

[(Prev) (Sen) + (1 - Prev) (1 - Spec)].

The gold standard refers to a definitive diagnosis obtained by biopsy, surgery, autopsy, long-term follow-up or another acknowledged standard. Within reason, the gold standard should have a Sen, Spec, +ve Pred values, -ve Pred values, and Accu of 100%. The ability of a test to discriminate between normal and abnormal individuals is described by its Sen and Spec. Sen and Spec are inversely related to each other, and can be altered by changing the reference interval or the normal range.

In other words, one can only be improved at the expense of the other. When a test has a Sen of 95% (5% false –ve) and a Spec of 95% (5% false +ve), for a disease with a 1% Prev, its +ve Pred value is only 16% but its –ve Pred value is 99%.

The relationship between Prev and +ve Pred value with a Sen of 95% are shown below:

Disease Prev (%)	+ve Pred value (%)
0	0
0.1	2
1	16
5	50
10	68
20	83
50	95

75	98
99	99.9
100	100

There are different definitions of "normal" such as Gaussian (Normal), Percentile, Risk factor, Culturally desirable, Diagnostic, and Therapeutic – none of them are perfect and each has different properties and consequences. An inevitable consequence arises as follows: when using the Percentile, if the normal range includes the lower 95% of diagnostic test results, then the likelihood that a given patient will be called normal when subjected to this test is 95%. If this same patient undergoes n independent diagnostic tests (independent in the sense that they are probing totally different organs or functions), the likelihood that the patient will be called normal is now 0.95 raised to the power of n. As an example, for n = 35 tests, the patient has only 0.95^{35} or 16.6% of being called normal. The reference interval (RI) for a quantitative test with a Gaussian distribution is defined as the Mean +/- 2 SD, namely 2.5% to 97.5% which encompasses approximately 95% of the results, where SD is the Standard Deviation. (Note that 2 SD is more accurately equal to 95.5%.) Inevitably 5% of entirely normal people will have test results outside the reference interval. Unfortunately, most analytes do not have a Gaussian distribution, generally being skewed towards the higher values – this often causes the yield of a negative value for the lower limit of the RI. Another example is that of Thyroid Stimulating Hormone (TSH) where its reference values are portrayed as logarithmically distributed. Depending on the method used, the RI is often quoted as 0.4 – 5.0 mIU/L, so that 'expected mean' is (0.4+5)/2 or 2.7 mIU/L. However, the actual mean and median values are about 1 mIU/L. Some methods used

119

to alleviate this problem include using logarithm, normal or log-normal probability graph and non-parametric method.

Before a clinician can draw conclusions on a patient's test result he or she usually compares it with the reference interval provided, and often also with the result previously obtained from the same patient. In so doing, the clinician needs to be aware of the effect on the result of some or all the following: analytical variation, biological variation, drugs & disease processes, and prevalence of the disease considered. The reference interval given for a quantitative test result must be appropriate for the patient e.g. physiological factors such as gender, race, age, posture (erect/supine), activity, pregnancy, etc must be taken into consideration if applicable.

Conclusions on test results can be used in one or more ways, for example, to assist a diagnosis, assess the progress of a disease process, determine the extent of dysfunction, assess the effect of treatment, monitor the stability of function, estimate a risk factor, as screening tests for "health", case finding, for research purposes, and so on.

When we talk about the validity and reliability of test results, we need to talk about factors affecting test results. These can be divided into: -

Pre-analytical variables – includes physiological factors, stress, diet, biological variation, drugs, medical history, patient preparation, specimen collection and transport.

Analytical variables – includes precision and accuracy of test method and factors which may interfere with a particular assay. These give rise to the 4 characteristics of analytical methods namely specificity, interference, bias & imprecision which clinical users should be aware.

Post-analytical variables – includes data entry & calculations by laboratory staff, result validation, interpretation of the result, data transfer and the method used to report the results.

Quality control and Quality assurance are the two processes used to control the impact of these variables.

Biological variation and Reference interval

It is crucial to understand the link between biological variation and reference interval. All physiological parameters have dynamic biological variations which are of a continuous nature. However, the monitoring of our many bodily physiological parameters (with different diverse properties) such as a substance concentration or activity, heart rate, heart or brain electrical activities, etc can be on an intermittent (discrete) or continuous basis.

In practice, the observed total (actual) variation of a performed 'Test Y' on 'Patient X' reflects the sum effects of biological (physiological) variation, and pre-analytical, analytical & post-analytical variables. Stated in another way: Total variation = Biological variation + Analytical variation, where Analytical (or Test) variation refers to pre-analytical, analytical & post-analytical variables. (Note that here we are dealing only with the within-person biological variation.) Let us talk about one of the characteristics from the analytical variables, namely analytical imprecision. Imprecision stem from either the sum effects of small changes in the performance of the instrumentation (within-run imprecision), or the sum effects of small changes due to different operators, different batches of reagents, instrument maintenance, etc (between-run or day-to-day imprecision). Test Y is to be performed twice on each of the blood samples collected from Patient X on different days. The difference between the two results for each day is calculated. Then, the within-run imprecision of the test assay (SD_A, standard deviation) is as follows:

121

$$SD_A = \sqrt{\left(\dfrac{\text{sum of } (differences)^2}{2 \text{ X number of pairs}}\right)}.$$

Analytical or test imprecision is quoted either as the value for 1 SD_A, or as the coefficient of variation [CV; where CV = 100 (1 SD_A / Mean) % at the mean value which the SD_A was determined]. A good test will have a small CV relative to the absolute value, whilst a poor test will have much wider limits. In real life, a blood sample from Patient X will only have Test Y performed once (and not reported as a mean of several estimations) - therefore, a result for Test Y means that there is a 95% probability that this result lies in the range of +/-2 SD_A. Next, when Test Y is performed on Patient X on two blood samples obtained from two different occasions, the standard error for the two results is $\sqrt{(2\ SDA^2)} = 1.4\ SD_A$. For 95% probability that two results on Patient X are different, they must differ by more than 2.8 SD_A. Furthermore if the known effect of biological variation SD_B is also taken into account, then total variation $SD_T = \sqrt{(SDA^2 + SDB^2)}$. Similarly, for 95% probability that two results on Patient X are different, they must differ by more than 2.8 SD_T.

We have already referred to variation (SD) and variance (SD^2) above. Then $SDB^2 = SDT^2 - SDA^2$ or $SD_B = \sqrt{(SDT^2 - SDA^2)}$. However, for most of the purpose of our discussion, we can safely ignore the effects from the pre-analytical, analytical & post-analytical variables as they can be theoretically assumed either to be totally absent or to be of a constant magnitude. In other words, SD_B is approximately taken to be the SD_T, and for 95% probability that two results on Patient X are different, they must differ by more than 2.8 SD_B. Tests can be roughly divided into two groups in terms of

122

their "performance characteristics". This is depicted as below:

Test T1: Large intra-individual variation relative to the inter-individual variation.
The intra-individual variation is about the same as the inter-individual variation. If Test T1 is performed on certain analytes (e.g. plasma sodium or potassium), and it follows that the set-point and the within-person biological variation is about the same for everyone; then the population reference interval is about the same as the biological variation (plus analytical variation) for the individual.

Test T2: Small intra-individual variation relative to the inter-individual variation.
The intra-individual variation is much smaller than the inter-individual variation. On the other hand, for Test T2 performed on certain other analytes (e.g. plasma alkaline phosphatase activity), and it follows that the within-person set points are quite different between individuals as are the variations around the set-points; then the population reference interval is much wider than the biological variation (plus analytical variation) for the individual.

The mean value of the daily test results is termed the set-point for Patient X. As depicted above, the within-person (intra-individual) biological variation for a test may be either of a similar magnitude, or be much smaller, to the population (inter-individual) biological variation (or reference interval). A pathological change in Test T1 analytes will obviously be more readily detected than in Test T2 analytes.

The two main types of analytes' biological variations characteristics also roughly correlate with that of the two types of test performance characteristics.

A simple layman's explanation of the terms Science, Religion and Human body may run as follows:

(1) Science is the study of structure and behavior of physical and natural world and society, especially through observation and experiment. Particular areas of science include computer science, medical science, the science of engineering, etc. The physical sciences include chemistry, physics, etc. The social sciences include psychology, politics, etc. The natural sciences include botany, marine science, zoology, etc. The applied sciences include engineering, computing, etc. Science fiction is a type of writing based on imagined scientific discoveries of the future and often dealing with space travel, life on other planets, and so on.

(2) The human body is best analyzed through study on human physiology and anatomy. Physiology can be defined as science which treats the functions (and the physical and chemical processes involved) of living organism and its parts, and of a species or class of organism and their parts. Its goal is to explain those physical and chemical factors that are responsible for the origin, development, and progression of life. Therefore, physiological studies can be performed on acellular (without enveloping plasma membrane) organisms such as viruses and virus-like agents such as viroids, plasmids, prions, retrons, satellite nucleic acids, satellite viruses; unicellular (single-celled) organisms such as bacteria and amoebae; and multicellular (many-celled) organisms such as plants and animals (including human beings). Small wonders that studies on physiology can be conducted for all living organisms, plants and animals in this world – thus, we have viral physiology, bacterial physiology, cellular physiology, plant physiology, human physiology, and many more subdivisions. Comparative physiology is a study of organ functions in various types of animals, vertebrate and invertebrate, in an effort to find fundamental relations in the physiology of members of the entire animal kingdom.

Human physiology is the study of the functions of our body (or how our body works) through its mechanisms of action explained in terms of cause-and-effect sequences of physical and chemical processes. With the mechanistic approach, which is employed by physiologists, the mechanisms of action (the "how" of events that occur in the body) are emphasized. With the teleological approach, phenomena that occur in the body are explained in terms of their particular purpose in fulfilling a bodily need (the "why" of body processes), without considering how this outcome is accomplished. Physiology is closely interrelated with anatomy, the study of the structure of our body. The structure and function of the human body are inseparable; and to tell the story of how the body works, we must also provide sufficient anatomical background on the function of the body part being discussed.

Medicine is the art and science of the prevention and cure of disease. A person taking medicines is usually taken to imply that he or she is taking one or more drugs to prevent, control or cure a disease or illness. Health is the condition of the body or the mind. Health also refers to the state of being well and free from illness. Pathology is the science of diseases. Obviously, then the study of diseases will involve the study of pathophysiology.

(3) Religion involves the belief in the existence of a supernatural ruling power, the creator and controller of the universe, who has given to man and woman a spiritual nature which continues to exist after death of the body. The great religions of this world with various systems of faith and worship based on such belief include Christianity, Islam, Buddhism, and Hinduism.

One of my frequent worries in life is about being falsely influenced by, or wasting precious time in, reading scientific materials (be it journal articles, literature reviews, magazines, books, etc) from any fields that may be "inaccurate" or "imperfect" due to its underlying mathematical language, research and theories being not quite up-to-date or being superseded by more advanced and correct theories. But our current understanding of science in the 21st Century, based on the work and visions of our scientific forebear and current scientists, has provided the framework to enable us to unravel the important basics or fabrics of science – the knowledge of which will help us in gauging the "soundness" of scientific materials; and perhaps, more significantly, in allowing a person to read any scientific literature and watch or listen to any scientific materials with less apprehension and greater comprehension on what was on offer.

During our journey through this book, we have touched base with many aspects of basic (and sometimes more advanced) mathematics and science, and we have also occasionally explore uncharted territories and sometimes expect the unexpected! However, this exercise would have been a futile one had we not link them together with "common threads of network" (such as Complexity, Chaos and Fractals) underlying the "fabrics of science".

Using scientific aspect and non-scientific aspect criteria, this popular science book can be roughly separated into two parts. The scientific aspect of this book is about "Basics of Science" and "Human body". The Basics of Science is based mainly on relatively new 'Science of Complexity', and along the way, requiring all-important **Alphabet and Language of Science** [intended] to be exposed. These important aspects – to be the subject of my future Book Series – are centered on two innovative models: Spherical Model of Science and Spherical Model of Numbers.

The brief non-scientific aspect of this book is about Basics of Religion based mainly on nature and role of religion in our lives, and along the way, some of the controversies, conflicts and parallels of religion with science exposed. The terminology of "non-scientific aspect criteria" with respect to religion refers to more traditional view that there is nothing scientific about study or practice of various religions – whether true or false, many people will, of course, beg to disagree with this traditional assessment.

The study of religions is mainly based on historical, or origins, science. This type of science deals with the past, is limited because we cannot do experiments directly on past events and as history cannot be repeated, inferences require a deal of guesswork. In contrast to process, or operational, science; inferences and conclusions are closely related to experiments with little room for speculation. It involve doing experiments in

126

the present, making inferences from these results and then carrying out further experiments to test those ideas or hypothesis. This second type of science has given us many valuable advances in knowledge that have given us many wonderful things and has benefited mankind: landing men on the moon, modern medicine, cheap food, electricity, computers, and so on.

EPILOGUE

I am literally a self-taught mathematician. I reckon my back-up mathematical career pathway should by now be well established after writing up my two landmark research papers in 2019 on successfully solving Incompletely Predictable problems of Riemann hypothesis, Polignac's and Twin prime conjectures. However, I have to humbly wait for general public and wider scientific community opinion on this point. In any event, I am now content with the legacy of my life achievements in mathematics. Does the coined jargon Sexy Mathematics in this book *cheekily* mean that I have bragging rights to be regarded as a Sexy Mathematician? Probably not but I can always dream this to be true. Trust you have enjoy reading this book and gain something out of it. I hope I will have more time in the very near future writing my planned BOOK SERIES 2 depicting 'Alphabet and Language of Science' and more "colorful" benefits arising out of solving Riemann hypothesis, Polignac's and Twin prime conjectures with again incorporating Medical Materials.

Bibliography

Farzana Mitra, Shahead Chowdhury, Mike Shelley and Gary Williams (January 15, 2013). A Feasibility Study of Transdermal Buprenorphine Versus Transdermal Fentanyl in the Long-Term Management of Persistent Non-Cancer Pain. Townsville, Australia: Pain Medicine, Volume 14, Issue 1, Pages 75–83 https://doi.org/10.1111/pme.12011.

Furstenberg, H. (1955). On the infinitude of primes. http://dx.doi.org/10.2307/2307043. USA: Amer. Math. Monthly, 62, (5) 353.

Lago P, T. C. (2008). Lago P, TiozRemifentanil for percutaneous intravenous central catheter placement in preterm infant: a randomized controlled trial. . Italy: Ped Anesth 2008; 18: 736 – 744 https://onlinelibrary.wiley.com/doi/abs/10.1111/j.1460-9592.2008.02636.x.

Noe, T. (November 23, 2004). A100967 https://oeis.org/A100967. USA: The On-Line Encyclopedia of Integer Sequences.

Saidak, F. (2006). A New Proof of Euclid's theorem. http://dx.doi.org/10.2307/27642094. USA: Amer. Math. Monthly, 113, (10) 937.

Sloane, N. J. (1964). A000001 https://oeis.org/A000001 (formerly published as M0098 N0035. Number of groups of order n.). USA: The On-Line Encyclopedia of Integer Sequences.

Ting, J. (2016). Key Role of Dimensional Analysis Homogeneity in Proving Riemann Hypothesis and Providing Explanations on the Closely Related Gram Points https://doi.org/10.5539/jmr.v8n4p1. Brisbane: Journal of Mathematics Research.

Ting, J. (2016). Rigorous Proof for Riemann Hypothesis Using the Novel Sigma-power Laws and Concepts from the Hybrid Method of Integer Sequence Classification https://doi.org/10.5539/jmr.v8n3p9. Brisbane: Journal of Mathematics Research.

Ting, J. (April 12, 2019). Solving Incompletely Predictable problem Riemann hypothesis with Dirichlet Sigma-Power Law http://vixra.org/pdf/1903.0483v6.pdf. Brisbane: viXra.

Ting, J. (April 26, 2019). Solving Incompletely Predictable problems Polignac's and Twin Prime conjectures using Information-Complexity conservation http://vixra.org/pdf/1904.0214v3.pdf. Brisbane: viXra.

Ting, J. (August 15, 2013). Hybrid integer sequence A228186
https://oeis.org/A228186. Brisbane: The On-Line Encyclopedia of Integer
Sequences.

Ting, J. Y. (2012). Supramaximal elevation in B-type natriuretic peptide and its N-
terminal fragment levels in anephric patients with heart failure: a case series
https://doi.org/10.1186/1752-1947-6-351. Sydney: Journal of medical case
reports.

Zhang, Y. (2014). Bounded gaps between primes.
http://dx.doi.org/10.4007/annals.2014.179.3.7. USA: Ann. Math. 179(3)
(2014) 1121 – 1174.

Solving Incompletely Predictable problem Riemann hypothesis with Dirichlet Sigma-Power Law

John Y. C. Ting

Published in viXra on April 12 2019

Abstract Riemann hypothesis proposed all nontrivial zeros to be located on critical line of Riemann zeta function. Treated as Incompletely Predictable problem, we obtain Dirichlet Sigma-Power Law as final proof of solving this problem. This Law is derived as equation and inequation from original Dirichlet eta function (proxy function for Riemann zeta function). Performing a parallel procedure help explain closely related Gram points.

Keywords Dimensional analysis, Dirichlet Sigma-Power Law, Gram points, Incompletely Predictable problems, Inequation, Nontrivial zeros, Riemann hypothesis

2010 Mathematics Subject Classification. 11A41, 11M26

1 Introduction

As Incompletely Predictable entities, Gram and virtual Gram points are dependently calculated using *complex* equation Riemann zeta function, $\zeta(s)$, or its proxy Dirichlet eta function, $\eta(s)$, in critical strip (denoted by $0 < \sigma < 1$). Gram[y=0], Gram[x=0] and Gram[x=0,y=0] points respectively refer to x-axis, y-axis and Origin intercepts at critical line (denoted by $\sigma = \frac{1}{2}$). Gram[y=0] and Gram[x=0,y=0] points are respectively synonymous with traditional *'Gram points'* and *nontrivial zeros* with the former further discussed in Segment A2, Appendix A.

131

Virtual Gram[y=0] and virtual Gram[x=0] points respectively refer to x-axis and y-axis intercepts at non-critical lines (denoted by $\sigma \neq \frac{1}{2}$). Virtual Gram[x=0,y=0] points do not exist. Activities to prove associated open problem in number theory Riemann hypothesis and explain Gram[y=0] and Gram[x=0] points equate to solving Incompletely Predictable problems. Claims from these activities are only meaningful when provided with definitions for relevant terms in Segment A1, Appendix A. Dependently calculated using *complex algorithm* Sieve of Eratosthenes, prime and composite numbers as Incompletely Predictable numbers are also depicted there.

In increasing size, arbitrary Set **X** can be countable finite set (CFS), countable infinite set (CIS) or uncountable infinite set (UIS). Cardinality of Set **X**, $|\mathbf{X}|$, measures the "number of elements" in Set **X**. E.g. Set **negative Gram[y=0] point** has CFS of negative Gram[y=0] point with $|$**negative Gram[y=0] point**$|$ = 1, Set **N** has CIS of natural numbers with $|\mathbf{N}| = \aleph_0$, and Set **R** has UIS of real numbers with $|\mathbf{R}| = c$ (cardinality of the continuum).

$$\zeta(s) = \frac{e^{\left(\ln(2\pi) - 1 - \frac{\gamma}{2}\right)}}{2(s-1)\Gamma\left(1 + \frac{s}{2}\right)}\Pi_\rho\left(1 - \frac{s}{\rho}\right)e^\rho = \pi^{\frac{s}{2}}\frac{\Pi_\rho\left(1 - \frac{s}{\rho}\right)}{2(s-1)\Gamma\left(1 + \frac{s}{2}\right)}$$

Proposed in 1859 by German mathematician Bernhard Riemann (September 17, 1826 – July 20, 1866), Riemann hypothesis is mathematical statement on $\zeta(s)$ that critical line denoted by $\sigma = \frac{1}{2}$ contains complete Set **nontrivial zeros** with $|$**nontrivial zeros**$| = \aleph_0$. Alternatively, this hypothesis is geometrical statement on $\zeta(s)$ that generated curves when $\sigma = \frac{1}{2}$ contain complete Set **Origin intercepts** with $|$**Origin intercepts**$| = \aleph_0$. Depicted in full and abbreviated version, Hadamard product is infinite product expansion of $\zeta(s)$ based on

Weierstrass's factorization theorem displaying a simple pole at s = 1. It contains both trivial and nontrivial zeros indicating their common origin from $\zeta(s)$. Set **trivial zeros** occurs at σ = -2, -4, -6, -8, -10,..., ∞ with |**trivial zeros**| = $\aleph_0$ due to Γ function term in denominator. Nontrivial zeros occur at s = ϱ with γ denoting Euler-Mascheroni constant.

Remark 1.1. Computationally checking for first 10,000,000,000,000 nontrivial zeros location on critical line implies but does not prove Riemann hypothesis to be true.

Locations of first 10,000,000,000,000 nontrivial zeros on critical line are previously confirmed to be correct. Hardy in 1914[1] and Hardy and Littlewood in 1921[2] showed infinite nontrivial zeros on critical line by considering moments of certain functions related to $\zeta(s)$. This discovery cannot constitute rigorous proof for Riemann hypothesis because they have not exclude theoretical existence of nontrivial zeros located away from this line.

Remark 1.2. We can apply useful concepts from exact and inexact Dimensional analysis homogeneity to well-defined equations and inequations.

Respectively for 'base quantities' such as *length, mass* and *time*; their fundamental SI 'units of measurement' meter (m) is defined as distance travelled by light in vacuum for time interval 1/299 792 458 s with speed of light c = 299,792,458 ms^{-1}, kilogram (kg) is defined by taking fixed numerical value Planck constant h to be 6.626 070 15 X 10^{-34} Joules·second (Js) [whereby Js is equal to kgm^{2}s^{-1}] and second (s) is defined in terms of $\Delta v Cs = \Delta(^{133}Cs)_{hfs} = 9,192,631,770$ s^{-1}. Derived SI units such as J and ms^{-1} respectively represent 'base quantities' *energy* and *velocity*. The word 'dimension' is commonly used to indicate all those mentioned 'units of measurement' in well-defined equations.

Dimensional analysis (DA) is an analytic tool with DA homogeneity and non-homogeneity (respectively) denoting valid and invalid equation occurring when 'units of measurements' for 'base quantities' are "balanced" and "unbalanced" across both sides of the equation. E.g. equation 2 m + 3 m = 5 m is valid and

133

equation 2 m + 3 kg = 5 mkg is invalid (respectively) manifesting DA homogeneity and non-homogeneity.

Let (2n) and (2n-1) be 'base quantities' in Dirichlet Sigma-Power Laws formatted in simplest forms as equations and inequations. E.g. DA on exponent $\frac{1}{2}$ in $(2n)^{\frac{1}{2}}$ in simplest form is correct but DA on exponent $\frac{1}{4}$ in equivalent $(2^2 n^2)^{\frac{1}{4}}$ *not* in simplest form is incorrect. Fractional exponents as 'units of measurement' given by $(1 - \sigma)$ for equations and $(\sigma + 1)$ for inequations when $\sigma = \frac{1}{2}$ coincide with exact DA homogeneity[1]; and $(1 - \sigma)$ for equations and $(\sigma + 1)$ for inequations when $\sigma \neq \frac{1}{2}$ coincide with inexact DA homogeneity[2].

Respectively for equations and inequations, exact DA homogeneity at $\sigma = \frac{1}{2}$ denotes $\sum$(all fractional exponents) as $2(1-\sigma)$ and $2(\sigma + 1)$ equates to ["exact"] whole number '1' and '3'; and inexact DA homogeneity at $\sigma \neq \frac{1}{2}$ denotes $\sum$(all fractional exponents) as $2(1-\sigma)$ and $2(\sigma + 1)$ equates to ["inexact"] fractional number '$\neq$ 1' and '$\neq$ 3'.

Footnote 1, 2: Exact and inexact DA homogeneity occur in Dirichlet Sigma-Power Laws as equations or inequations for Gram[y=0] points, Gram[x=0] points and nontrivial zeros. *Law of Continuity* is a heuristic principle *whatever succeed for the finite, also succeed for the infinite*. Then these Laws which inherently manifest themselves on finite and infinite time scale should "succeed for the finite, also succeed for the infinite".

Outline of proof for Riemann hypothesis. To simultaneously satisfy two mutually inclusive conditions: I. *With rigid manifestation of exact DA homogeneity*, Set nontrivial zeros with |nontrivial zeros| = $\aleph_0$ is located on critical line (viz. $\sigma = \frac{1}{2}$)

134

when $2(1 - \sigma)$ [or $2(\sigma + 1)$] as $\sum$(all fractional exponents) = whole number '1' [or '3'] in Dirichlet Sigma-Power Law[3] as equation [or inequation]. II. *With rigid manifestation of inexact DA homogeneity*, Set nontrivial zeros with |nontrivial zeros| = $\aleph_0$ is not located on non-critical lines (viz. $\sigma \neq \frac{1}{2}$) when $2(1-\sigma)$ [or $2(\sigma +1)$] as $\sum$(all fractional exponents) = fractional number '$\neq$ 1' [or '$\neq$ 3'] in Dirichlet Sigma-Power Law[3] as equation [or inequation].

Footnote 3: Derived from original $\eta(s)$ (*proxy* for $\zeta(s)$) as equation or inequation, this Law symbolizes end-result proof on Riemann hypothesis.

Riemann hypothesis mathematical foot-prints. Six identifiable steps to prove Riemann hypothesis: *Step 1* Use η(s), *proxy* for ζ(s), in critical strip. *Step 2* Apply Euler formula to η(s). *Step 3* Obtain "simplified" Dirichlet eta function which intrinsically incorporates *actual location [but not actual positions]* of all nontrivial zeros[4]. *Step 4* Apply Riemann integral to "simplified" Dirichlet eta function in discrete (summation) format. *Step 5* Obtain Dirichlet Sigma-Power Law in continuous (integral) format as equation or inequation. *Step 6* Note exact and inexact DA homogeneity on their fractional exponents.

Footnote 4: Respectively Gram[y=0] points, Gram[x=0] points and nontrivial zeros are Incompletely Predictable entities with actual positions determined by setting $\sum \text{Im}\{\eta(s)\} = 0$, $\sum \text{Re}\{\eta(s)\} = 0$ and $\sum \text{ReIm}\{\eta(s)\} = 0$ to *dependently* calculate relevant positions of all preceding entities in neighborhood. Respectively actual location of Gram[y=0] points, Gram[x=0] points and nontrivial zeros; and virtual Gram[y=0] points, virtual Gram[x=0] points and "absent" nontrivial zeros occur precisely at $\sigma = \frac{1}{2}$ and $\sigma \neq \frac{1}{2}$.

2 Riemann zeta and Dirichlet eta functions

L-functions form an integral part of 'L-functions and Modular Forms Database' (LMFDB) with far-reaching implications. In perspective, ζ(s) is simplest example of an L-function. ζ(s) is a function of complex variable s (= $\sigma \pm$

135

it) that analytically continues sum of infinite series $\zeta(s) = \sum_{n=1}^{\infty} \frac{1}{n^s} =$

$\frac{1}{1^s} + \frac{1}{2^s} + \frac{1}{3^s} + \ldots.$ The common convention is to write s as $\sigma + it$ with $i = \sqrt{-1}$,

and σ and t real. Valid for $\sigma > 0$, we write $\zeta(s)$ as $\text{Re}\{\zeta(s)\} + i\cdot\text{Im}\{\zeta(s)\}$ and note

that $\zeta(\sigma + it)$ when $0 < t < +\infty$ is the complex conjugate of $\zeta(\sigma - it)$ when $-\infty < t < 0$.

Also known as alternating zeta function, $\eta(s)$ must act as *proxy* for $\zeta(s)$ in

critical strip (viz. $0 < \sigma < 1$) containing critical line (viz. $\sigma = \frac{1}{2}$) because $\zeta(s)$ only

converges when $\sigma > 1$. This implies $\zeta(s)$ is undefined to left of this region in

critical strip which then requires $\eta(s)$ representation instead. They are related to

each other as $\zeta(s) = \gamma \cdot \eta(s)$ with proportionality factor $\gamma = \frac{1}{(1 - 2^{1-s})}$ and

$\eta(s) = \sum_{n=1}^{\infty} \frac{(-1)^{n+1}}{n^s} = \frac{1}{1^s} - \frac{1}{2^s} + \frac{1}{3^s} + \ldots.$

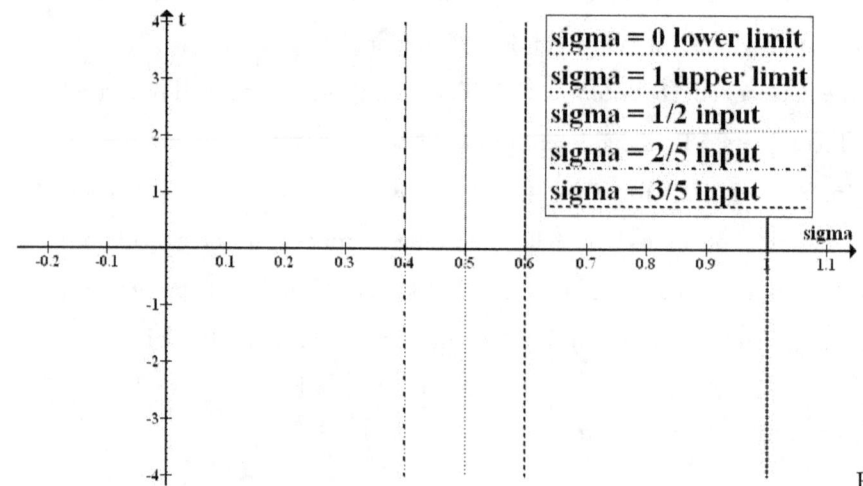

g. 1 INPUT for $\sigma = \frac{1}{2}, \frac{2}{5}, and \frac{3}{5}$. $\zeta(s)$ has CIS of Completely Predictable trivial zeros at σ = all negative even numbers and CIS of Incompletely Predictable

nontrivial zeros at $\sigma = \dfrac{1}{2}$ for various t values.

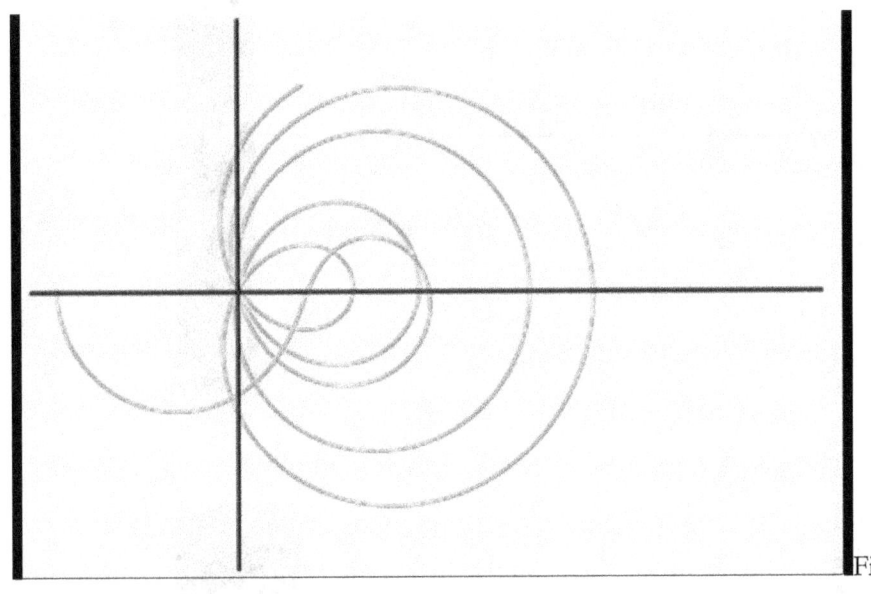

g. 2 OUTPUT for $\sigma = \dfrac{1}{2}$. Schematically depicted polar graph of $\zeta(\dfrac{1}{2}+\imath t)$ plotted along critical line for real values of t running from 0 to 34, horizontal axis: as $Re\{\zeta(\dfrac{1}{2}+\imath t)\}$, and vertical axis: $Im\{\zeta(\dfrac{1}{2}+\imath t)\}$. Note presence of Origin intercepts which are totally absent in Figures 3 and 4 [with identical axes definitions].

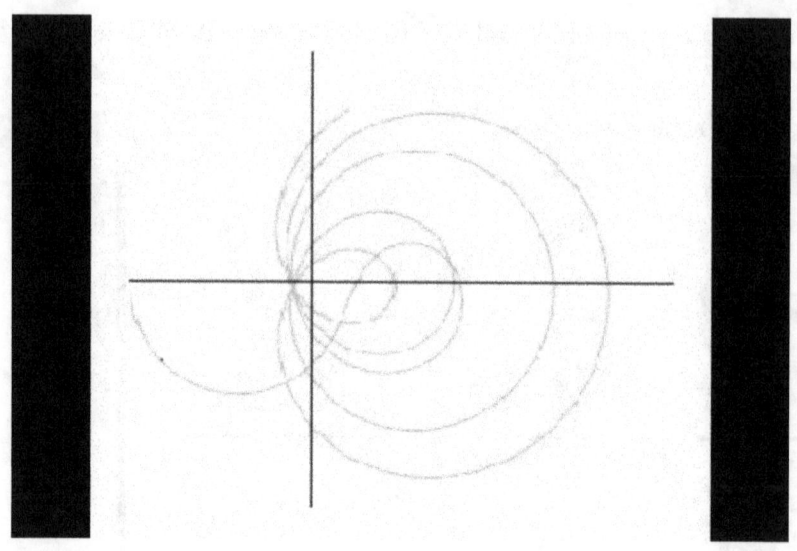

Fig. 3 OUTPUT for $\sigma = \dfrac{2}{5}$.

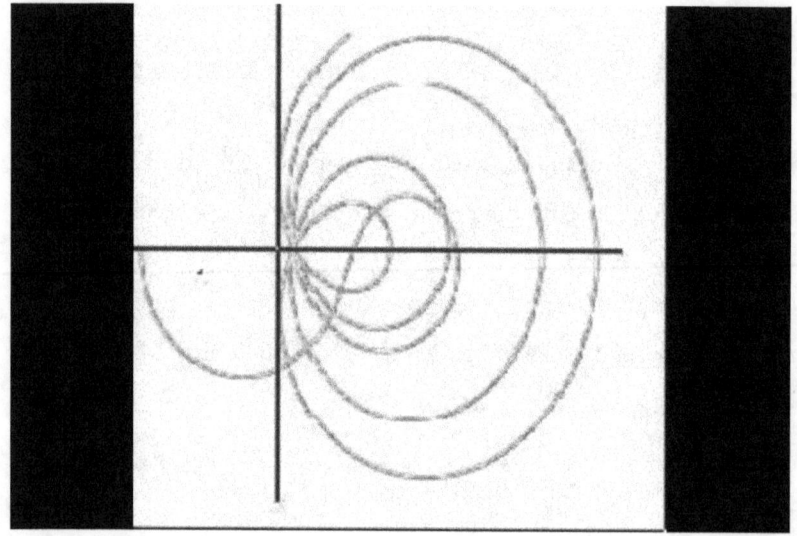

Fig. 4 OUTPUT for $\sigma = \dfrac{3}{5}$.

$$\zeta(s) = \sum_{n=1}^{\infty}\frac{1}{n^s} \tag{1}$$

$$= \frac{1}{1^s} + \frac{1}{2^s} + \frac{1}{3^s} + \dots$$

$$= \Pi\text{p } \textit{prime } \frac{1}{(1-p^{-s})}$$

$$= \frac{1}{(1-2^{-s})} \cdot \frac{1}{(1-3^{-s})} \cdot \frac{1}{(1-5^{-s})} \cdot \frac{1}{(1-7^{-s})} \cdot \frac{1}{(1-11^{-s})} ... \frac{1}{(1-p^{-s})} ...$$

Eq. (1) is defined for only $1 < \sigma < \infty$ region where $\zeta(s)$ is absolutely convergent. There are no zeros located here. In Eq. (1), equivalent Euler product formula with product over prime numbers [instead of summation over natural numbers] can also represent $\zeta(s)$.

$$\zeta(s) = 2^s \pi^{s-1} \sin\left(\frac{\pi s}{2}\right) \cdot \Gamma(1-s) \cdot \zeta(1-s) \qquad (2)$$

With $\sigma = \frac{1}{2}$ as symmetry line of reflection, Eq. (2) is Riemann's functional equation valid for $-\infty < \sigma < \infty$. It can be used to find all trivial zeros on horizontal line at $\pi s = 0$ occurring πs when $\sigma = $ -2, -4, -6, -8, -10,..., ∞ whereby $\zeta(s)$ = 0 because factor $\sin\left(\frac{\pi s}{2}\right)$ vanishes. Γ is gamma function, an extension of factorial function [a product function denoted by ! notation whereby n! = n(n−1)(n−2)...(n−(n−1))] with its argument shifted down by 1, to real and complex numbers. That is, if n is a positive integer, $\Gamma(n) = (n−1)!$

$$\zeta(s) = \frac{1}{(1-2^{1-s})} \sum_{n=1}^{\infty} \frac{(-1)^{n+1}}{n^s}$$

$$= \frac{1}{(1-2^{1-s})} \left(\frac{1}{1^s} - \frac{1}{2^s} + \frac{1}{3^s} + ...\right) \qquad (3)$$

Eq. (3) is defined for all $\sigma > 0$ values except for simple pole at $\sigma = 1$. As alluded to above, $\zeta(s)$ without $\frac{1}{(1-2^{1-s})}$ viz. $\sum_{n=1}^{\infty} \frac{(-1)^{n+1}}{n^s}$ is $\eta(s)$. It is a holomorphic function of s defined by analytic continuation and is mathematically defined at $\sigma = 1$ whereby

139

analogous trivial zeros with presence only for $\eta(s)$ [but not for $\zeta(s)$] on vertical straight

line $\sigma = 1$ are found at $s = 1 \pm i.\dfrac{2\pi k}{ln\boxed{}(2)}$ where k = 1, 2, 3, 4, 5, ..., ∞.

Figure 1 depict complex variable s (= $\sigma \pm$ it) as INPUT with x-axis denoting real part Re{s} equating to σ; and y-axis denoting imaginary part Im{s} equating to t. Figures 2, 3 and 4 respectively depict $\zeta(s)$ as OUTPUT for real values of t running from 0 to 34 at for $\sigma = \dfrac{1}{2}$ (critical line), for $\sigma = \dfrac{2}{5}$ (non-critical line), and for $\sigma = \dfrac{3}{5}$ (non-critical line) with x-axis denoting real part Re{$\zeta(s)$} and y-axis denoting imaginary part Im{$\zeta(s)$}. There are infinite types-of-spirals possibilities associated with each σ value arising from all infinite σ values in critical strip. Mathematically proving all nontrivial zeros location on critical line as denoted by solitary for $\sigma = \dfrac{1}{2}$ value equates to geometrically proving all Origin intercepts occurrence at solitary $\sigma = \dfrac{1}{2}$ value. Both result in rigorous proof for Riemann hypothesis.

3 Prerequisite lemma, corollary and propositions for Riemann hypothesis

Original equation $\eta(s)$, *proxy* for $\zeta(s)$, is treated as unique mathematical object with key properties and behaviors. Containing all x-axis, y-axis and Origin intercepts, it will intrinsically incorporate *actual location [but not actual positions]* of all Gram[y=0] points, Gram[x=0] points and nontrivial zeros. Proofs on lemma, corollary and propositions on nontrivial zeros depict exact and inexact DA homogeneity in both derived equation and inequation. Parallel procedure on Gram[y=0] and Gram[x=0] points in Appendix B depict exact and inexact DA homogeneity in similarly derived equations and inequations.

140

Lemma 3.1. "Simplified" Dirichlet eta function is derived directly from Dirichlet eta function with Euler formula application and it will intrinsically incorporate actual location [but not actual positions] of all nontrivial zeros.

Proof. Denote complex number (**C**) as $z = x + i \cdot y$. Then $z = \mathrm{Re}(z) + i \cdot \mathrm{Im}(z)$ with $\mathrm{Re}(z) = x$ and $\mathrm{Im}(z) = y$; modulus of z, $|z| = \sqrt{(\mathrm{Re}(z)^2 + \mathrm{Im}(z)^2)} = \sqrt{(x^2 + y^2)}$); and $|z|^2 = x^2 + y^2$.

Euler formula is commonly stated as $e^{ix} = \cos x + i \cdot \sin x$. Euler identity (where $x = \pi$) is $e^{i\pi} = \cos \pi + i \cdot \sin \pi = -1 + 0$ [or stated as $e^{i\pi} + 1 = 0$]. The n^s of $\zeta(s)$ is expanded to $n^s = n^{(\sigma + it)} = n^\sigma e^{\ln(n) \cdot i}$ since $n^t = e^{t \ln(n)}$. Apply Euler formula to n^s result in $n^s = n^\sigma(\cos(t \ln(n)) + i \cdot \sin(t \ln(n))$. This is written in trigonometric form [designated by shorthand notation n^s(Euler)] whereby n^σ is modulus and $t \ln(n)$ is polar angle (argument). Apply n^s(Euler) to Eq. (1). Then

$$\zeta(s) = \mathrm{Re}\{\zeta(s)\} + i \cdot \mathrm{Im}\{\zeta(s)\} \text{ with } \mathrm{Re}\{\zeta(s)\} = \sum_{n=1}^{\infty} n^{-\sigma}\cos(t \ln(n)) \quad \text{and } \mathrm{Im}\{\zeta(s)\}$$

$$= \sum_{n=1}^{\infty} n^{-\sigma}\sin(t \ln(n))$$
. As Eq. (1) is defined only for $\sigma > 1$ where zeros never occur, we will not carry out further treatment here.

Apply n^s(Euler) to Eq. (3). Then $\zeta(s) = \gamma \cdot \eta(s) = \gamma \cdot [\mathrm{Re}\{\eta(s)\} + i \cdot \mathrm{Im}\{\eta(s)\}]$ with

$$\mathrm{Re}\{\eta(s)\} = \sum_{n=1}^{\infty}((2n-1)^{-\sigma}\cos(t \ln(2n-1)) - (2n)^{-\sigma}\cos(t \ln(2n)))$$
;

$$\mathrm{Im}\{\eta(s)\} = \sum_{n=1}^{\infty}((2n)^{-\sigma}\sin(t\ln(2n)) - (2n-1)^{-\sigma}\sin(t\ln(2n)))$$
;

and proportionality factor $\gamma = \dfrac{1}{(1 - 2^{1-s})}$.

Complex number s in critical strip is designated by $s = \sigma + it$ for $0 < t < +\infty$ and $s = \sigma - it$ for $-\infty < t < 0$. Nontrivial zeros equating to $\zeta(s) = 0$ give rise to our desired $\eta(s) = 0$. Modulus of $\eta(s)$, $|\eta(s)|$, is defined as $\sqrt{(\mathrm{Re}\{\eta(s)\})^2}$

141

$+(\text{Im}\{\eta(s)\})^2)$ with $|\eta(s)|^2 = (\text{Re}\{\eta(s)\})^2 +(\text{Im}\{\eta(s)\})^2$. Mathematically $|\eta(s)| = |\eta(s)|^2 = 0$ is an unique condition giving rise to $\eta(s) = 0$ occurring only when $\text{Re}\{\eta(s)\} = \text{Im}\{\eta(s)\} = 0$ as any non-zero values for $\text{Re}\{\eta(s)\}$ and/or $\text{Im}\{\eta(s)\}$ will always result in $|\eta(s)|$ and $|\eta(s)|^2$ having nonzero values. Important implication is that sum of $\text{Re}\{\eta(s)\}$ and $\text{Im}\{\eta(s)\}$ equating to zero [given by Eq. (4)] must always hold when $|\eta(s)| = |\eta(s)|^2 = 0$ and consequently $\eta(s) = 0$.

$$\sum \text{ReIm}\{\eta(s)\} = \text{Re}\{\eta(s)\}+\text{Im}\{\eta(s)\} = 0 \qquad (4)$$

In principle, advocating for existence of theoretical s values leading to non-zero values in $\text{Re}\{\eta(s)\}$ and $\text{Im}\{\eta(s)\}$ depicted as possibility $+\text{Re}\{\eta(s)\} = -\text{Im}\{\eta(s)\}$ or $-\text{Re}\{\eta(s)\} = +\text{Im}\{\eta(s)\}$ could satisfy Eq. (4). This reverse implication is not necessarily true as these s values will not result in $|\eta(s)| = |\eta(s)|^2 = 0$. In any event, we need not consider these two possibilities since solving Riemann hypothesis involves nontrivial zeros defined by $\eta(s) = 0$ with non-zero values in $\text{Re}\{\eta(s)\}$ and/or $\text{Im}\{\eta(s)\}$ being not compatible with $\eta(s) = 0$.

Riemann hypothesis proposed all nontrivial zeros to be located on critical line. This location is conjectured to be uniquely associated with presence of exact DA homogeneity in derived equation and inequation of Dirichlet Sigma-Power Law with Eq. (4) intrinsically incorporated into this Law as the $\eta(s) = 0$ definition for nontrivial zeros equates to Eq. (4).

Apply trigonometry identity $\cos(x) - \sin(x) = \sqrt{2}\sin(x + \frac{3}{4}\pi)$ to $\text{Re}\{\eta(s)\}+\text{Im}\{\eta(s)\}$ to get Eq. (5) with terms in last line built by mixture of terms from $\text{Re}\{\eta(s)\}$ and $\text{Im}\{\eta(s)\}$. $\sum \text{ReIm}\{\eta(s)\}$

$$= \sum_{n=1}^{\infty} [((2n-1)^{-\sigma}\cos(t\ln(2n-1)) - (2n-1)^{-\sigma}\sin(t\ln(2n-1))$$

$$- (2n)^{-\sigma}\cos(t\ln(2n)) + (2n)^{-\sigma}\sin(t\ln(2n))]$$

$$= \sum_{n=1}^{\infty} [((2n-1)^{-\sigma}\sqrt{2}\sin(t\ln(2n-1) + \frac{3}{4}\pi) - (2n)^{-\sigma}\sqrt{2}\sin\left(t\ln(2n) + \frac{3}{4}\pi\right)]$$

$$(5)$$

142

When depicted in terms of Eq. (4), Eq. (5) becomes

$$\sum_{n=1}^{\infty}(2n)^{-\sigma}\sqrt{2}\sin\left(t\,ln\boxed{\cdot}(2n)+\frac{3}{4}\pi\right)=\sum_{n=1}^{\infty}(2n-1)^{-\sigma}\sqrt{2}\sin(t\,ln\boxed{\cdot}(2n-1)+\frac{3}{4}\pi)$$

$$\sum_{n=1}^{\infty}(2n)^{-\sigma}\sqrt{2}\sin\left(t\,ln\boxed{\cdot}(2n)+\frac{3}{4}\pi\right)-\sum_{n=1}^{\infty}(2n-1)^{-\sigma}\sqrt{2}\sin\left(t\,ln\boxed{\cdot}(2n-1)+\frac{3}{4}\pi\right)$$

$$=0 \qquad\qquad (6)$$

Eq. (6) in discrete (summation) format is a non-Hybrid integer sequence equation – see Appendix C. $\eta(s)$ calculations for all σ values result in infinitely many non-Hybrid integer sequence equations for $0 < \sigma < 1$ critical strip region of interest with n = 1, 2, 3, 4, 5,..., ∞ as discrete integer number values, or n = 1 to ∞ as continuous real numbers values with Riemann integral application. These equations will geometrically represent entire plane of critical strip, thus (at least) allowing our proposed proof to be of a complete nature.

Eq. (6) being the "simplified" Dirichlet eta function derived directly from $\eta(s)$ will intrinsically incorporate *actual location [but not actual positions]* of all nontrivial zeros. *The proof is now complete for Lemma 3.1□.*

Proposition 3.2. Dirichlet Sigma-Power Law in continuous (integral) format given as equation and inequation can both be derived directly from "simplified" Dirichlet eta function in discrete (summation) format with Riemann integral application.

Proof. In Calculus, integration is reverse process of differentiation viewed geometrically as area enclosed by curve of function and x-axis. Apply definite integral I between points a and b is to compute its value when $\Delta x \to 0$, i.e. I =

$\lim_{\Delta x \to 0} \Sigma\ f(x_i)\Delta x_i = \int_{a}^{b} f(x)dx$. This is Riemann integral of function f(x) in interval [a, b] where a<b. Apply Riemann integral to "simplified" Dirichlet eta function in $[\Delta x \to 1]$ discrete (summation) format which intrinsically incorporates *actual location [but not actual positions]* of all nontrivial zeros criterion to

obtain Dirichlet Sigma-Power Law in [$\Delta x \to 0$] continuous (integral) format with the later validly representing the former. Then Dirichlet Sigma-Power Law will also fullfil this criterion. Due to resemblance to power law functions in σ from s $= \sigma + it$ being exponent of a power function n^σ, logarithm scale use, and harmonic $\zeta(s)$ series connection in Zipf's law; we elect to call this Law by its given name. A characteristic and crucial step of this Law is its exact formula expression in usual mathematical language [$y = f(x_1, x_2)$ format description for a 2-variable function with ($2n$) and ($2n-1$) parameters] consist of $y = f(t, \sigma)$ with discrete $n = 1, 2, 3, 4, 5,..., \infty$ or continuous $n = 1$ to ∞; $-\infty < t < +\infty$; and $0 < \sigma$ <1.

With steps of manual integration shown using indefinite integrals [for simplicity], solve definite integral below based on numerator portion of R1 with ($2n$) parameter in Eq. (6):

$$\int_1^\infty \frac{2^{\frac{1}{2}-\sigma} \sin\left(t\, \ln\boxed{}(2n) + \frac{3}{4}\pi\right)}{n^\sigma}\, dn = \int_1^\infty -\frac{\sin\left(t\ln(2n)\right) - \cos\left(t\ln(2n)\right)}{2^\sigma n^\sigma}\, dn.$$ We

deduce most other important integrals to be "variations" of this particular integral containing (i) deletion of $(2n)^{-\sigma}$, $\sqrt{2}$ or $\frac{3}{4}^{-\pi}$ terms, and/or (ii) interchange of sine and cosine function. We check all derived antiderivatives to be correct using computer algebra system Maxima.

Simplifying and applying linearity, we obtain
$$2^{\frac{1}{2}-\sigma}\int \frac{\sin\left(t\, \ln\boxed{}(2n) + \frac{3}{4}\pi\right)}{n^\sigma}\, dn.$$

Now solving $\displaystyle\int \frac{\sin\left(t\,\ln(2n) + \frac{3}{4}\pi\right)}{n^{\sigma}}$ dn. Substitute $u = t\,\ln(2n) + \frac{3}{4}\overset{-\pi}{} \rightarrow$ dn =

$\dfrac{n}{t}$du, use $n^{1-\sigma} = e$ $\dfrac{(1-\sigma)\,(u - t\ln(2n) - \frac{3}{4}\pi)}{t}$ =

$\dfrac{e^{\frac{(\sigma-1)\,(4t\ln(2n) - 3\pi)}{4t}}}{t} \displaystyle\int e^{\frac{(1-\sigma)\,u}{t}}\,\sin(u)\,du.$

Now solving $\displaystyle\int e^{\frac{(1-\sigma)\,u}{t}}\,\sin(u)\,du$. We integrate by parts twice in a row:

$\displaystyle\int f\,g' = f\,g - \int f'\,g.$

First time: $f = \sin(u)$, $g' = e^{\frac{(1-\sigma)\,u}{t}}$.

Then $f' =$

cos (u)

$g =$

$(1-\sigma)te^{\frac{(1-\sigma)\,u}{t}}$
$\overline{\sigma^2 - 2\sigma + 1}$

$= \dfrac{(1-\sigma)te^{\frac{(1-\sigma)\,u}{t}}\,\sin(u)}{\sigma^2 - 2\sigma + 1} - \displaystyle\int \dfrac{(1-\sigma)te^{\frac{(1-\sigma)\,u}{t}}\,\cos(u)}{\sigma^2 - 2\sigma + 1}du$

Second time: $f =$

cos(u)

$g' =$

$$\boxed{\dfrac{(1-\sigma)te^{\frac{(1-\sigma)u}{t}}}{\sigma^2-2\sigma+1}}$$

Then $f' = -\sin(u)$, $g = \dfrac{t^2 e^{\frac{(1-\sigma)u}{t}}}{\sigma^2-2\sigma+1}$:

$$= \dfrac{(1-\sigma)te^{\frac{(1-\sigma)u}{t}}\sin(u)}{\sigma^2-2\sigma+1} - \left(\dfrac{t^2 e^{\frac{(1-\sigma)u}{t}}\cos(u)}{\sigma^2-2\sigma+1} - \int -\dfrac{t^2 e^{\frac{(1-\sigma)u}{t}}\sin(u)}{\sigma^2-2\sigma+1}du \right)$$

Apply linearity:

$$= \dfrac{(1-\sigma)te^{\frac{(1-\sigma)u}{t}}\sin(u)}{\sigma^2-2\sigma+1} - \left(\dfrac{t^2 e^{\frac{(1-\sigma)u}{t}}\cos(u)}{\sigma^2-2\sigma+1} + \right.$$

$$\left. \dfrac{t^2}{\sigma^2-2\sigma+1}\int e^{\frac{(1-\sigma)u}{t}}\sin(u)du \right)$$

As integral $\displaystyle\int e^{\frac{(1-\sigma)u}{t}}\sin(u)du$ appears again on Right Hand Side, we can solve for it:

$$= \dfrac{\dfrac{(1-\sigma)e^{\frac{(1-\sigma)u}{t}}\sin(u)}{t} - e^{\frac{(1-\sigma)u}{t}}\cos(u)}{\dfrac{\sigma^2-2\sigma+1}{t^2}+1}$$

Plug in solved integrals:

$$= \dfrac{e^{\frac{(\sigma-1)4t\ln(2)+3\pi}{4t}}}{4t}\int e^{\frac{(1-\sigma)u}{t}}\sin(u)du$$

$$= \frac{e^{\frac{(\sigma-1)4t\ln(2)+3\pi}{4t}}\left(\dfrac{(1-\sigma)e^{\frac{(1-\sigma)u}{t}}\sin(u)}{t} - e^{\frac{(1-\sigma)u}{t}}\cos(u)\right)}{\left(\dfrac{\sigma^2-2\sigma+1}{t^2}+1\right)t}$$

Undo substitution u = $\dfrac{t\ln(2n)-\dfrac{3}{4}\pi}{}$ and simplifying:

$$= \frac{e^{\frac{(\sigma-1)4t\ln(2)+3\pi}{4t}}\left(\dfrac{(1-\sigma)e^{\frac{(\sigma-1)(t\ln(2)+\frac{3}{4}\pi)}{t}}\sin(t\ln(2)+\frac{3}{4}\pi)}{t} - e^{\frac{(\sigma-1)(t\ln(2)+\frac{3}{4}\pi)}{t}}\cos\left(t\ln(2)+\frac{3}{4}\pi\right)\right)}{\left(\dfrac{\sigma^2-2\sigma+1}{t^2}+1\right)t}$$

Plug in solved integrals:

$$2^{\frac{1}{2}-\sigma}\int \frac{\sin\left(t\, \ln(2n)+\frac{3}{4}\pi\right)}{n^\sigma}\ dn$$

$$= \frac{2^{\frac{1}{2}-\sigma}\, e^{\frac{(\sigma-1)4t\ln(2)+3\pi}{4t}}\left(\dfrac{(1-\sigma)e^{\frac{(\sigma-1)(t\ln(2)+\frac{3}{4}\pi)}{t}}\sin(t\ln(2)+\frac{3}{4}\pi)}{t} - e^{\frac{(\sigma-1)(t\ln(2)+\frac{3}{4}\pi)}{t}}\cos\left(t\ln(2)+\frac{3}{4}\pi\right)\right)}{\left(\dfrac{\sigma^2-2\sigma+1}{t^2}+1\right)t}$$

By rewriting and simplifying, $\displaystyle\int_1^\infty \frac{2^{\frac{1}{2}-\sigma}\sin\left(t\, \ln(2n)+\frac{3}{4}\pi\right)}{n^\sigma}\,dn$ is finally solved as

$$\left[\frac{(2n)^{1-\sigma}\left((t+\sigma-1)\sin(t\ln(2n))+(t-\sigma+1)\cos(t\ln(2n))\right)}{2(t^2+(\sigma-1)^2)}+C\right]_1^\infty \qquad (7)$$

For denominator portion of R1 with (2n-1) parameter in Eq. (6), Eq. (7) equates to

$$\left[\frac{(2n-1)^{1-\sigma}((t+\sigma-1)sin(t\,ln(2n-1))+(t-\sigma+1)cos(t\,ln(2n-1)))}{2(t^2+(\sigma-1)^2)}+C\right]_1^\infty \tag{8}$$

Dirichlet Sigma-Power Law as equation derived from Eq. (6) is given by:

$$\frac{1}{2\left(t^2+(\sigma-1)^2\right)}\cdot[(2n)^{1-\sigma}((t+\sigma-1)\sin(t\,\ln(2n))+(t-\sigma+1)\cos(t\,\ln(2n)))-$$

$$(2n-1)^{1-\sigma}((t+\sigma-1)\sin(t\,\ln(2n-1))+(t-\sigma+1)\cos(t\,\ln(2n-1)))]_1^\infty=0 \quad (9)$$

Apply Ratio Study to Eq. (6) – see Segment A3, Appendix A. This involves [intentional] incorrect but "balanced" rearrangement of terms in Eq. (6) giving rise to Eq. (10) which is a non-Hybrid integer sequence inequation. Left-hand side contains 'cyclical' sine function in first term (Ratio R1) and 'non-cyclical' power function in second term (Ratio R2).

$$\frac{\sqrt{2}sin\left(t\,ln⬚(2n)+\frac{3}{4}\pi\right)}{\sqrt{2}sin\left(t\,ln⬚(2n-1)+\frac{3}{4}\pi\right)}-\frac{(2n)^{-\sigma}}{(2n-1)^{-\sigma}}\neq 0$$

$$\tag{10}$$

Apply Riemann integral to selected parts of Eq. (10) without depicting steps of calculation:

$$\int_1^\infty \sqrt{2}sin\left(t\,ln⬚(2n)+\frac{3}{4}\pi\right)dn =$$

$$\left[\frac{(2n)((t-1)sin(t\,ln(2n))+(t+1)cos(t\,ln(2n)))}{2(t^2+1)}+C\right]_1^\infty \quad \text{and}$$

$$\int_1^\infty \sqrt{2}sin\left(t\,ln⬚(2n-1)+\frac{3}{4}\pi\right)dn =$$

$$\left[\frac{(2n-1)((t-1)sin(t\,ln(2n-1))+(t+1)cos(t\,ln(2n-1)))}{2(t^2+1)}+C\right]_1^\infty .$$

$$\int_1^\infty (2n)^\sigma dn = \left[\frac{(2n)^{\sigma+1}}{2(\sigma+1)} + c\right]_1^\infty$$ and

$$\int_1^\infty (2n-1)^\sigma dn = \left[\frac{(2n-1)^{\sigma+1}}{2(\sigma+1)} + c\right]_1^\infty.$$

Dirichlet Sigma-Power Law as inequation derived from Eq. (10) is given by:

$$\left[\frac{(2n)((t-1)\sin(t\,\ln(2n)) + (t+1)\cos(t\,\ln(2n)))}{(2n-1)((t-1)\sin(t\,\ln(2n-1)) + (t+1)\cos(t\,\ln(2n-1)))} - \frac{(2n)^{\sigma+1}}{(2n-1)^{\sigma+1}}\right]_1^\infty$$

$$\neq 0 \qquad\qquad (11)$$

Intended derivation of Dirichlet Sigma-Power Law as equation and inequation have been successful. *The proof is now complete for Proposition 3.2□.*

Proposition 3.3. Exact Dimensional analysis homogeneity at $\sigma = \frac{1}{2}$ in Dirichlet SigmaPower Law as equation and inequation is (respectively) indicated by $\sum$(all fractional exponents) = whole number '1' and '3'.

Proof. Dirichlet Sigma-Power Law as equation for $\sigma = \frac{1}{2}$ value is given by:

$$\frac{1}{2+\frac{1}{2}} \cdot \left[(2n)^{\frac{1}{2}}\left((t-\frac{1}{2})\sin(t\,\ln(2n)) + (t+\frac{1}{2})\cos(t\,\ln(2n))\right) -\right.$$

$$\left. (2n-1)^{\frac{1}{2}}\left((t-\frac{1}{2})\sin(t\,\ln(2n-1)) + (t+\frac{1}{2})\cos(t\,\ln(2n-1))\right)\right]_1^\infty =$$

$$= 0 \qquad\qquad (12)$$

Respectively evaluation of definite integrals Eq. (12), Eq. (24) and Eq. (26) using limit as n $\rightarrow$ +∞ for $0 < t < +\infty$ enable countless computations resulting in t values for CIS of nontrivial zeros, Gram[y=0] points and Gram[x=0] points. We illustrate this for Eq. (12) as expanded antiderivative [depicted as linear combination of sine and cosine waves: a sin x + b cos x = c sin (x + φ) with c = $\sqrt{(a^2 + b^2)}$ and φ = atan2 (b, a) = $tan^{-1}(\frac{b}{a})$ for a > 0].

149

$$(2\infty)^{\frac{1}{2}} \sin((t\,\ln 2\infty)) + \tan^{-1}\left(\frac{t+\frac{1}{2}}{t-\frac{1}{2}}\right) - (2\infty-1)^{\frac{1}{2}} \sin((t\,\ln 2\infty\text{-}1)) + \tan^{-1}\left(\frac{t+\frac{1}{2}}{t-\frac{1}{2}}\right) -$$

$$2^{\frac{1}{2}} \sin((t\,\ln 2)) + \tan^{-1}\left(\frac{t+\frac{1}{2}}{t-\frac{1}{2}}\right) + \frac{t+\frac{1}{2}}{2t^2+\frac{1}{2}} = 0$$

(2∞) and $(2\infty$ - 1) involve exponent $\frac{1}{2}$, sin and ln functions. At relevant t values for all nontrivial zeros, (first term - second term) = (- third term + fourth term).

Dirichlet Sigma-Power Law as inequation for $\sigma = \frac{1}{2}$ value is given by:

$$\left[\frac{(2n)((t-1)\sin(t\,\ln(2n)) + (t+1)\cos(t\,\ln(2n)))}{(2n-1)((t-1)\sin(t\,\ln(2n-1)) + (t+1)\cos(t\,\ln(2n-1)))} - \frac{(2n)^{\frac{1}{2}}}{(2n-1)^{\frac{1}{2}}}\right]_{\infty 1} \neq$$

0   (13)

$\sum$(all fractional exponents) as $2(1 - \sigma)$ = whole number '1' for Eq. (12) and $2(\sigma + 1)$ = whole number '3' for Eq. (13). These findings signify presence of complete set nontrivial zeros for Eq. (12) and Eq. (13). *The proof is now complete for Proposition 3.3□.*

Corollary 3.4. Inexact Dimensional analysis homogeneity at $\sigma \neq \frac{1}{2}$ [illustrated using $\sigma = \frac{2}{5}$] in Dirichlet Sigma-Power Law as equation and inequation is (respectively) indicated by $\sum$(all fractional exponents) = fractional number '$\neq$ 1' and '$\neq$ 3'.

Proof. Dirichlet Sigma-Power Law as equation for $\sigma = \frac{2}{5}$ value is given by:

150

$$\frac{1}{2+\frac{18}{25}} \cdot [(2n)^{\frac{3}{5}} \left((t-\frac{3}{5})\sin(t\ln(2n)) + (t+\frac{3}{5})\cos(t\ln(2n)) \right) -$$

$$(2n-1)^{\frac{3}{5}} \left((t-\frac{3}{5})\sin(t\ln(2n-1)) + (t+\frac{3}{5})\cos(t\ln(2n-1)) \right)]_1^\infty = 0 \quad (14)$$

Dirichlet Sigma-Power Law as inequation for $\sigma = \dfrac{2}{5}$ value is given by:

$$\left[\frac{(2n)((t-1)\sin(t\,ln(2n)) + (t+1)\cos(t\,ln(2n)))}{(2n-1)((t-1)\sin(t\,ln(2n-1)) + (t+1)\cos(t\,ln(2n-1)))} - \frac{(2n)^{\frac{7}{5}}}{(2n-1)^{\frac{7}{5}}} \right]_1^\infty \neq 0$$

(15)

$\sum$(all fractional exponents) as $2(1 - \sigma)$ = fractional number '$\neq$ 1' for Eq. (14) and $2(\sigma + 1)$ = fractional number '$\neq$ 3' for Eq. (15). These findings signify absence of complete set nontrivial zeros for Eq. (14) and Eq. (15). The proof is now complete for Corollary 3.4□.

4 Rigorous proof for Riemann hypothesis summarized as Theorem Riemann I – IV

$$\zeta(s) = \frac{1}{s-1} + \frac{1}{2} + 2 \int_0^\infty \frac{\sin(s\arctan t)}{(1+t^2)^{\frac{1}{2}}(e^{2\pi t} - 1)} dt$$

is integral relation (cf. Abel Plana summation formula[3], [4]) for all s $\in$ C and s $\neq$ 1. This integral is insufficient for our purpose as it involves integration w.r.t. t [instead of n] for $\zeta(s)$ [instead of $\eta(s)$]. Rigorous proof for Riemann hypothesis is summarized by Theorem Riemann I – IV. One could obtain this proof with only using Dirichlet Sigma-Power Law [solely] as equation. For completeness and clarification of this proof, we supply following important mathematical arguments.

For $0 < \sigma < 1$, then $0 < 2(1-\sigma) < 2$. The only whole number between 0 and 2 is '1' which coincide with $\sigma = \dfrac{1}{2}$. When $0 < \sigma < \dfrac{1}{2}$ and $\dfrac{1}{2} < \sigma < 1$, then $0 < 2(1-\sigma) < 1$ and $1 < 2(1-\sigma) < 2$.

For $0 < \sigma < 1$, $2 < 2(\sigma + 1) < 4$. The only whole number between 2 and 4 is

'3' which coincide with $\sigma = \dfrac{1}{2}$. When $0 < \sigma < \dfrac{1}{2}$ and $\dfrac{1}{2} < \sigma < 1$, then $2 < 2(\sigma +1)$ < 3 and $3 < 2(\sigma +1) < 4$.

Legend: R = all real numbers. For $0 < \sigma < 1$, σ consist of $0 < R < 1$. For $0 < 2(1-\sigma) < 2$ and $2 < 2(\sigma +1) < 4$, $2(1-\sigma)$ and $2(\sigma +1)$ must (respectively) consist of $0 < R < 2$ and $2 < R < 4$. An important caveat is that previously used phrases such as "fractional exponent σ" and "$\sum$(all fractional exponents) = whole number '1' [or '3'] and fractional number '$\neq$ 1' [or '$\neq$ 3']", although not incorrect *per se*, should respectively be replaced by "real number exponent σ" and "$\sum$(all real number exponents) = whole number '1' [or '3'] and real number '$\neq$ 1' [or '$\neq$ 3'][5]" for complete accuracy. We apply this caveat to Theorem Riemann I – IV.

Footnote 5: As whole numbers $\subset$ real numbers, one could also depict this phrase as "$\sum$(all real number exponents) = real number '1' [or '3'] and real number '$\neq$ 1' [or '$\neq$ 3']".

Theorem Riemann I. Derived from *proxy* Dirichlet eta function, "simplified" Dirichlet eta function will exclusively contain *de novo* property for actual location [but not actual positions] of all nontrivial zeros.

Proof. The phrase "actual location [but not actual positions] of all nontrivial zeros" can be validly shortened to "actual location of all nontrivial zeros" as used in Theorem Riemann II, III and IV. *The proof for Theorem Riemann I is now complete as it successfully incorporates proof for Lemma 3.1□.*

Theorem Riemann II. Dirichlet Sigma-Power Law [in continuous (integral) format] as equation and inequation which are both derived from "simplified" Dirichlet eta function [in discrete (summation) format] will exclusively manifest exact DA homogeneity in equation and inequation only when real number exponent $\sigma = \dfrac{1}{2}$.

Proof. *The proof for Theorem Riemann II is now complete as it successfully incorporates proofs from Proposition 3.2 on derivation for equation and inequation of Dirichlet Sigma-Power Law [with both containing de novo property for "actual location of all nontrivial zeros"] and Proposition 3.3 on manifestation of exact DA homogeneity in Dirichlet Sigma-Power Law as equation and inequation when real number exponent $\sigma = \frac{1}{2}$ □.*

Theorem Riemann III. Real number exponent $\sigma = \frac{1}{2}$ in Dirichlet Sigma-Power Law as equation and inequation satisfying exact DA homogeneity is identical to σ variable in Riemann hypothesis which propose σ to also have exclusive value of $\frac{1}{2}$ (representing critical line) for "actual location of all nontrivial zeros", thus fully supporting Riemann hypothesis to be true with further clarification by Theorem Riemann IV.

Proof. Since $s = \sigma \pm \imath t$, complete set of nontrivial zeros which is defined by $\eta(s) = 0$ is exclusively associated with one (and only one) particular $\eta(\sigma \pm \imath t) = 0$ value solution, and by default one (and only one) particular σ [conjecturally] $= \frac{1}{2}$ value solution. When performing exact DA homogeneity on Dirichlet Sigma-Power Law as equation and inequation [with both containing *de novo* property for "actual location of all nontrivial zeros"], the phrase "If real number exponent σ has exclusively $\frac{1}{2}$ value, only then will exact DA homogeneity be satisfied" implies one (and only one) possible mathematical solution. Theorem Riemann III reflect Theorem Riemann II on presence of exact DA homogeneity for $\sigma = \frac{1}{2}$ in Dirichlet Sigma-Power Law as equation and inequation. This Law has identical σ variable as that referred to by Riemann hypothesis [whereby σ here uniquely refer

153

to critical line]. *The proof for Theorem Riemann III is now complete as it independently refers to simultaneous association of confirmed (i) solitary* $\sigma = \frac{1}{2}$ *value in Dirichlet Sigma-Power Law as equation and inequation satisfying exact DA homogeneity and (ii) critical line defined by solitary* $\sigma = \frac{1}{2}$ *value being the "actual location [but with no request to determine actual positions]" of all nontrivial zeros as proposed in original Riemann hypothesis*□.

Theorem Riemann IV. Condition 1. All $\sigma \neq \frac{1}{2}$ values (non-critical lines), viz. 0 $< \sigma < \frac{1}{2}$ and $\frac{1}{2} < \sigma < 1$ values, exclusively does not contain "actual location of all nontrivial zeros" [manifesting *de novo* inexact DA homogeneity in equation and inequation], together with Condition 2. One (and only one) $\sigma = \frac{1}{2}$ value (critical line) exclusively contains "actual location of all nontrivial zeros" [manifesting *de novo* exact DA homogeneity in equation and inequation], fully support Riemann hypothesis to be true when these two mutually inclusive conditions are met.

Proof. Condition 2 Theorem Riemann IV simply reflect proof from Theorem Riemann III [incorporating Proposition 3.3] for "actual location of all nontrivial zeros" exclusively on critical line manifesting *de novo* exact DA homogeneity $\sum$(all real number exponents) = whole number '1' for equation [or '3' for inequation]. *The proof for Condition 2 Theorem Riemann IV is now complete*□. Corollary 3.4 confirms *de novo* inexact DA homogeneity manifested as $\sum$(all real number exponents) = real number '$\neq$ 1' for equation [or '$\neq$ 3' for inequation] by all $\sigma \neq \frac{1}{2}$ values (non-critical lines) that are exclusively not associated with "actual location of all nontrivial zeros". Applying inclusion-exclusion principle: Exclusive presence of nontrivial zeros on critical line for Condition 2 Theorem Riemann IV implies

154

exclusive absence of nontrivial zeros on non-critical lines for Condition 1 Theorem Riemann IV. *The proof for Condition 1 Theorem Riemann IV is now complete*□.

We logically deduce that explicit mathematical explanation why presence and absence of nontrivial zeros[6] should (respectively) coincide precisely with $\sigma = \frac{1}{2}$ and $\sigma \neq \frac{1}{2}$ [literally the Completely Predictable meta-properties ('overall' *complex properties*)] will require "complex" mathematical arguments. Attempting to provide explicit mathematical explanation with "simple" mathematical arguments would intuitively mean nontrivial zeros have to be (incorrectly and impossibly) treated as Completely Predictable entities.

Footnote 6: Completely Predictable meta-properties for Gram and virtual Gram points equating to "Presence of Gram[y=0] and Gram[x=0] points, and virtual Gram[y=0] and virtual Gram[x=0] points (respectively) coincide precisely with $\sigma = \frac{1}{2}$, and $\sigma \neq \frac{1}{2}$".

5 Conclusions

In our Hybrid method of Integer Sequence classification, a formula is either non-Hybrid or Hybrid integer sequence. Inequation with two 'necessary' Ratio (R) or equation with one 'unnecessary' R contains non-Hybrid integer sequence. Equation with one 'necessary' R contains Hybrid integer sequence. "In the limit" Hybrid integer sequence approach unique Position X, it becomes non-Hybrid integer sequence for all Positions ≥ Position X.

Consider kinetic energy (KE) in MJ with m_o = rest mass in kg and v = velocity in m s^{-1}. In classical mechanics concerning low velocity with v << c, Newtonian KE = $\frac{1}{2}m_o v^2$. In relativistic mechanics concerning high velocity with v ≥ 0.01c, Relativistic

$$KE = \frac{m_o c^2}{\sqrt{(1-(v^2/c^2))}} - m_o c^2$$. Obtained from the later by binomial approximation or by taking first two terms of Taylor expansion for reciprocal square root, the former approximates the later well at low speed.

We arbitrarily divide DA homogeneity into inexact DA homogeneity for ["<100% accuracy"] Newtonian KE and exact DA homogeneity for ["100% accuracy"] Relativistic KE. "In the limit" ['<100% accuracy'] Newtonian KE at low speed approach ['100% accuracy'] Relativistic KE at high speed, we achieve *perfection*.

Analogy: "In the limit" all three version of Dirichlet Sigma-Power Laws for Gram[y=0] points, Gram[x=0] points and nontrivial zeros as *'<100% accuracy'* inequations approach *perfection* as *'100% accuracy'* equations, compliance with inexact DA homogeneity becomes compliance with exact DA homogeneity. We note R1 terms in all inequations contain (2n) and (2n-1) 'base quantities' but these are not endowed with fractional exponent $(\sigma+1)$ as relevant 'unit of measurement'. As Incompletely Predictable problems, we gave relatively elementary proof of Riemann hypothesis and explain closely related Gram points whereby various "meta-properties" such as exact and inexact DA homogeneity occur in (respectively) equations and inequations of relevant Dirichlet Sigma-Power Laws. Harnessed key benefit from successful proof for Riemann hypothesis is often stated as "With this one solution, we have proven five hundred theorems or more at once". This apply to important theorems in number theory that rely on properties of Riemann zeta function or Dirichlet eta function such as location of trivial and nontrivial zeros. E.g., we delineate prime number theorem by prime counting function $\pi(x)$ [which is defined as number of primes $\leq$ x].

156

Appendix A: Definitions and Supplementary materials

Exposition on definitions and related commentaries is crucial to help solve Riemann hypothesis and explain closely related Gram points as Incompletely Predictable problems.

Segment A1. *Completely Predictable and Incompletely Predictable numbers*

Completely Unpredictable numbers arising from totally chaotic physical processes give rise to countable infinite set (CIS) of measured true random numbers. In this sense, computational pseudorandom number generators using deterministic logic are not regarded as sources for true random numbers. Two types of Predictable numbers: CIS of Completely, and CIS of Incompletely Predictable numbers with former "contained" in *simple* equations or algorithms obeying '*Simple* Elementary Fundamental Laws', and later "contained" in *complex* equations or algorithms obeying '*Complex* Elementary Fundamental Laws'.

A Completely, and Incompletely Predictable number is locationally defined as a number whose position is *independently* determined by simple calculations using simple equation or algorithm without, and *dependently* by complex calculations using complex equation or algorithm with needing to know related positions of all preceding numbers in neighborhood. Both types of Predictable number exist as either rational [integers or fractions of two integers] numbers (Q) or irrational [algebraic or transcendental] numbers (R – Q). A well-defined set of R – Q will twice obey relevant location definition in CIS R – Q themselves and in CIS numerical digits after decimal point of each R – Q.

97 is an Incompletely Predictable number whose precise position is determined by computing positions of all preceding 24 prime numbers (P) using complex algorithm Sieve of Eratosthenes to conclude that 97 is 25th P. Calculated using simple algorithm, 97 is also [i = (97+1)/2] 49th odd number (O) which is a Completely Predictable number. 98 & 99 are respectively [i = 98/2] 49th even number (E) & [i = (99+1)/2] 50th O which are Completely Predictable numbers

157

calculated using simple algorithm. Determined indirectly using complex algorithm Sieve of Eratosthenes, 98 & 99 are respectively also 72nd & 73rd composite numbers (C) which are Incompletely Predictable numbers.

Computing Riemann zeta function (or specifically its *proxy* Dirichlet eta function) and Sieve of Eratosthenes will, respectively, supply Incompletely Predictable nontrivial zeros, Gram[y=0] & Gram[x=0] points and P & C. CIS of nontrivial zeros (denoted by imaginary part parameter t) = CIS of transcendental numbers = 14.134725, 21.022040, 25.010858, 30.424876, 32.935062, 37.586178,... [rounded off to six decimal places]. CIS of all P = Countable Finite Set (CFS) of all even P + CIS of all odd P = 2, 3, 5, 7, 11, 13,... whereby P '2' when treated as E is also regarded as a Completely Predictable number.

The three sets of nontrivial zeros, Gram[y=0] points and Gram[x=0] points, respectively, will *dependently* constitute three sets of Origin intercepts (or simultaneous x- & y-axes intercepts), x-axis intercepts and y-axis intercepts. Traditional 'Gram points' [see Segment A2 below] are x-axis intercepts with choice of index 'n' for 'Gram points' historically chosen such that first 'Gram point' corresponds to t value which is larger than (first) nontrivial zero located at t = 14.134725. By convention, first six Gram[y=0] points will occur with following values [rounded off to six decimal places]: at n = -3, t = 0; at n = -2, t = 3.436218; at n = -1, t = 9.666908; at n = 0, t = 17.845599; at n = 1, t = 23.170282; at n = 2, t = 27.670182.

The two sets of P 2, 3, 5, 7, 11, 13,... and C 4, 6, 8, 9, 10, 12,... will *dependently* constitute set of natural numbers (N) 1, 2, 3, 4, 5, 6,... minus first N '1'. Whole numbers (W) = N plus '0'. '0' & '1' are special numbers being neither P nor C as they represent nothingness (zero) and wholeness (one), and the idea of having factors for '0' & '1' is meaningless. Treating '0' & '1' here as Completely or Incompletely Predictable numbers is also meaningless.

CIS of numbers derived from well-defined simple/complex algorithms or equations are "dual numbers" displayed as Completely & Incompletely

Predictable number. Examples of Q '2' as P (& E), '97' as P (& O), '98' as C (& E) and '99' as C (& O) are described above. Examples of R – Q are described next. First & only negative Gram[y=0] point (by convention at n = -3) with Completely Predictable y = 0 value is obtained by substituting Completely Predictable t = 0 resulting in $\zeta(\overline{\tfrac{1}{2}} + \imath t) = \zeta(\overline{\tfrac{1}{2}})$ = -1.4603545, an Incompletely Predictable transcendental number [rounded off to seven decimal places] calculated as a limit similar to limit for Euler-Mascheroni constant or Euler gamma – its precise (1st) position can only be determined by computing positions of all preceding (nil) Gram[y=0] points in this case. With exception of this first Gram[y=0] point, all t values from Gram[y=0] points, Gram[x=0] points, and nontrivial zeros (Gram[x=0,y=0] points) are Incompletely Predictable transcendental numbers – these are respectively associated with Completely Predictable x = 0, y = 0, and simultaneous x = 0 & y = 0 values. First 'Gram point' (by convention at n = 0 & is associated with Completely Predictable x = 0 value from Incompletely Predictable t = 17.845599 substitution) is actually the 4th Gram[y=0] point whose precise (4th) position can only be determined by computing positions of all preceding (three) Gram[y=0] points in this case. First nontrivial zero associated with simultaneous x = 0 & y = 0 value [equating to ζ(s) = 0 whereby s = σ + $\imath t$ = $\overline{\tfrac{1}{2}}$ + $\imath t$] is Completely Predictable occurring with Incompletely Predictable t = 14.134725 value substitution – its precise (1st) position can only be determined by computing positions of all preceding (nil) nontrivial zeros in this case.

Remark A.1. Countable finite set (CFS) of Completely Predictable *simple properties* intrinsically present in simple equations or algorithms help us solve Completely Predictable problems containing countable infinite set (CIS) of Completely Predictable numbers; whereas CFS of Completely Predictable *complex properties* intrinsically present in complex equations or algorithms help us solve

Incompletely Predictable problems containing CIS of Incompletely Predictable numbers.

Simple properties are inferred from a phrase like: "...the simple equation or algorithm by itself will intrinsically incorporate actual location [and actual positions] of all Completely Predictable numbers". Solving Completely Predictable problems endowed with simple properties which are amendable to *simple* treatments using *usual* mathematical tools such as Calculus will result in their 'Simple Elementary Fundamental Laws'-based solutions. Complex properties are inferred from a phrase like: "...the complex equation or algorithm by itself will intrinsically incorporate actual location [but not actual positions] of all Incompletely Predictable numbers". Solving Incompletely Predictable problems endowed with complex properties which are amendable to *complex* treatments using *unusual* mathematical tools such as deriving complex equation Dirichlet Sigma-Power Law as well as using *usual* mathematical tools such as Calculus will result in their 'Complex Elementary Fundamental Laws'-based solutions.

Consider x for real number (R) values $\geq$ 1. Let y be Set R such that (simple equation) y = 2x or y = 2x - 1. This simple equation will "contain" the complete uncountable infinite set (UIS) of R [straight line of infinite length] commencing from Cartesian point (x=1, y=2) or (x=1, y=1). Computing y = 2x or y = 2x - 1 an infinite number of times – a *mathematical impasse* – will not *per se* result in its 'Simple Elementary Fundamental Laws'-based solution for gradient or slope = 2. This gradient (simple property) is obtained by trigonometrically calculating tangent of y = 2x or y = 2x - 1 straight line which = 2 or analyzing y = 2x or y = 2x - 1 equation using Differential Calculus viz. dy/dx = d(2x)/dx or d(2x-1)/dx = 2. Note: applying Integral Calculus from Fundamental Theorem of Calculus to

y = 2x or y = 2x - 1 for interval [1, +∞], viz. $\int_1^\infty (2x)dx$ or $\int_1^\infty (2x-1)dx$ =

$[x^2 + C]^\infty 1$ or $[x^2 + x + C]^\infty 1$ = (∞^2 +C) – (1^2 + C) or (∞^2–∞ + C) - (1^2 - 1 + C)

160

= ∞ result in 'Simple Elementary Fundamental Laws'-based solution for area (simple property) of infinite size enclosed by the straight line and x-axis.

Consider x≥1 integer number (Z) values for (simple algorithm) y = 2x or y = 2x - 1. We obtain "contained" complete Set E or Set O. Computing E or O infinitely often – a *mathematical impasse* – will not *per se* result in 'Simple Elementary Fundamental Laws'-based solution for gap between any two consecutive E (E gap) or O (O gap) will both = 2. This gradient-equivalent E gaps or O gaps (simple property) is obtained by transforming those algorithms from their "discrete" into "continuous" format [viz. "discrete" $\Delta x \rightarrow 1$ into "continuous" $\Delta x \rightarrow 0$] resulting in their gradients using either tangent method or Differential Calculus method. Then E or O gaps, both = 2, is numerically identical and mathematically equivalent to relevant gradients, both also = 2. Similar method of transforming from "discrete" into "continuous" format to help solve Riemann hypothesis involves applying Riemann integral to discrete-like equation of "simplified" Dirichlet eta function (in summation format) to obtain Dirichlet Sigma-Power Law [which is the continuous-like equation of "simplified" Dirichlet eta function (in integral format)].

Similar to Incompletely Predictable 'varying gaps' [equivalent to 'varying gradients'] between consecutive P (P gaps) & consecutive C (C gaps) [relevant to research on Polignac's and Twin prime conjectures], we have Incompletely Predictable 'varying gaps' [equivalent to 'varying gradients'] between consecutive nontrivial zeros (nontrivial zero gaps), consecutive Gram[y=0] points (Gram[y=0] points gaps) & consecutive Gram[x=0] points (Gram[x=0] points gaps). These 'varying gaps' or 'varying gradients' (complex properties) are geometrically related to different shapes/sizes of spirals depicted in Figure 2.

Segment A2. *Gram's Law and traditional 'Gram points'*

Named after Danish mathematician Jørgen Pedersen Gram (June 27, 1850 – April 29, 1916), traditional 'Gram points' (or Gram[y=0] points which are **x-axis intercepts** shown in figure above) are other conjugate pairs values on critical line defined by $\operatorname{Im}\{\zeta(\frac{1}{2} \pm \imath t)\} = 0$. They obey Gram's Rule and Rosser's Rule with interesting characteristic properties as outlined by our brief exposition below.

Z function is used to study Riemann zeta function on critical line. Defined in terms of Riemann-Siegel theta function & Riemann zeta function by $Z(t) = e^{\imath\theta(t)}\zeta(\frac{1}{2} + \imath t)$ whereby $\theta(t) = \arg(\Gamma(\frac{(2\imath t + 1)}{4})) - \frac{\ln\pi}{2} t$; it is also called Riemann-Siegel Z function, Riemann-Siegel zeta function, Hardy function, Hardy Z function, & Hardy zeta function.

The algorithm to compute $Z(t)$ is called Riemann-Siegel formula. Riemann zeta function on critical line, $\zeta(\frac{1}{2} + \imath t)$, will be real when $\sin(\theta(t)) = 0$. Positive real values of t where this occurs are called 'Gram points' and can also be described as points where $\frac{\theta(t)}{\pi}$ is an integer. Real part of this function on critical line tends to be positive, while imaginary part alternates more regularly between positive & negative values. That means sign of $Z(t)$ must be opposite to that of sine function most of the time, so one would expect nontrivial zeros of $Z(t)$ to alternate with zeros of sine term, i.e. when θ takes on integer multiples of π. This turns out to hold most of the time and is known as Gram's Rule (Law) – a law which is violated infinitely often though. Thus Gram's Law is statement that nontrivial zeros of $Z(t)$ alternate with 'Gram points'. 'Gram points' which satisfy Gram's Law are called 'good', while those that do not are called 'bad'. A Gram block is an interval such that its very first & last points are good 'Gram points' and all 'Gram points' inside this interval are bad. Counting nontrivial zeros then reduces to counting all 'Gram points' where Gram's Law is satisfied and adding the count of nontrivial zeros inside each Gram block. With this process we do not have to locate nontrivial zeros, and we just have to accurately compute $Z(t)$ to show that it changes sign.

162

Ratio Study and Inequations

A mathematical equation, containing one or more variables, is a statement that values of two ['left-hand side' (LHS) and 'right-hand side' (RHS)] mathematical expressions is related as equality: LHS = RHS; or as inequalities: LHS < RHS, LHS > RHS, LHS $\leq$ RHS, or LHS $\geq$ RHS. A ratio is one mathematical expression divided by another. The term 'unnecessary' Ratio (R) for any given equation is explained by two examples: (1) LHS = RHS and with rearrangement, 'unnecessary' R is given by $\frac{LHS}{RHS} = 1$ or $\frac{RHS}{LHS} = 1$; and (2) LHS > RHS and with rearrangement, 'unnecessary' R is given by $\frac{LHS}{RHS} > 1$ or $\frac{RHS}{LHS} < 1$.

Consider exponent y $\in$ all **R** values & base x $\in$ **R**$\geq$0 values for mathematical expression $\frac{x}{y}$. Equations such as $x^1 = x$, $x^0 = 1$ & $0^y = 0$ are all valid. Simultaneously letting both x & y = 0 is an incorrect mathematical action because xy as function of two-variables is not continuous & is thus undefined at Origin. But if we elect to intentionally carry out this "balanced" action [equally] on x & y, we obtain (simple) inequation $0^0 \neq 1$ with associated perpetual obeyance of '=' equality symbol in x^y for all applicable **R** values except when both x & y = 0. The Number '1' value in this inequation is justified by two arguments: I. Limit of x^y value as both x & y tend to zero (from right) is 1 [thus fully satisfying criterion "x^y is right continuous at the Origin"]; and II. Expression x^y is product of x with itself y times [and thus x^0, the "empty product", should be 1 (no matter what value is given to x)].

Mathematical operator 'summation' must obey the law: We can break up a summation across a sum or difference but not across a product or quotient viz, factoring a sum of quotients into a corresponding quotient of sums is an incorrect mathematical action. But if we elect to carry out this action equally on LHS & RHS products or quotients in a suitable equation, we obtain two (unique) 'necessary' R denoted by R1 for LHS and R2 for RHS whereby R1 ≠ R2 relationship will always hold. We define 'Ratio Study' as intentionally performing this incorrect [but "balanced"] mathematical action on suitable equation [equivalent to one (non-unique) 'unnecessary' R] to obtain its inequation [equivalent to two (unique) 'necessary' R]. Let C denote complex numbers. Set C is a field (but not an ordered field). Thus it is not possible to define a relation between two given (z1 & z2) C as z1 < z2 since inequality operation here is not compatible with addition and multiplication. But performing Ratio Study to obtain inequations involving C does not involve defining a relation between two C.

Appendix B: Prerequisite lemma, corollary and propositions for Gram[y=0] and Gram[x=0] points

For Gram[y=0] and Gram[x=0] points (and corresponding virtual Gram[y=0] and virtual Gram[x=0] points with totally different values), we apply a parallel procedure carried out on nontrivial zeros but only depict abbreviated treatments and discussions.

Lemma B.1. "Simplified" Gram[y=0] and Gram[x=0] points-Dirichlet eta functions are derived directly from Dirichlet eta function with Euler formula application and (respectively) they will intrinsically incorporate actual location [but not actual positions] of all Gram[y=0] and Gram[x=0] points.

Proof. For Gram[y=0] points, the equivalent of Eq. (4) and Eq. (6) are respectively given by Eq. (16) and Eq. (17) below.

$$\sum \text{ReIm}\{\eta(s)\} = \text{Re}\{\eta(s)\}+0, \text{ or simply } \text{Im}\{\eta(s)\} = 0 \quad (16)$$

For Gram[x=0] points, the equivalent of Eq. (4) and Eq. (6) are respectively given by Eq. (18) and Eq. (19) below.

$$\sum \text{ReIm}\{\eta(s)\} = 0+\text{Im}\{\eta(s)\}, \text{ or simply } \text{Re}\{\eta(s)\} = 0 \quad (18)$$

$$\sum_{n=1}^{\infty} (2n)^{-\sigma}cos(t\ ln▨(2n)) = \sum_{n=1}^{\infty} (2n-1)^{-\sigma}cos(t\ ln▨(2n-1))$$

$$\sum_{n=1}^{\infty}(2n)^{-\sigma}\cos(t\,\ln(2n)) \; - \; \sum_{n=1}^{\infty}(2n-1)^{-\sigma}\sin(t\,\ln(2n-1))$$
$$= 0$$

(19)

Eq. (17) and Eq. (19) being the "simplified" Gram[y=0] and Gram[x=0] points-Dirichlet eta functions derived directly from $\eta(s)$ will intrinsically incorporate *actual location [but not actual positions]* of (respectively) all Gram[y=0] and Gram[x=0] points. *The proof is now complete for Lemma B.1□.*

Proposition B.2. Gram[y=0] and Gram[x=0] points-Dirichlet Sigma-Power Laws in continuous (integral) format given as equations and inequations can both be (respectively) derived directly from "simplified" Gram[y=0] and Gram[x=0] points-Dirichlet eta functions in discrete (summation) format with Riemann integral application.

Proof. Antiderivatives below using (2n) parameter help obtain all subsequent equations: first two for Gram[y=0] points and second two for Gram[x=0] points.

$$\int_1^{\infty}(2n)^{-\sigma}\sin(t\ln(2n))dn = \left[-\frac{(2n)^{1-\sigma}((\sigma-1)\sin(t\ln(2n))+t\cos(t\ln(2n)))}{2\left(t^2+(\sigma-1)^2\right)}+C\right]_1^{\infty}$$

$$\int_1^{\infty}\sin(t\ln(2n))dn = \left[\frac{(2n)(\sin(t\ln(2n))-t\cos(t\ln(2n)))}{2(t^2+1)}+C\right]_1^{\infty}$$

$$\int_1^{\infty}(2n)^{-\sigma}\cos(t\ln(2n))dn = \left[\frac{(2n)^{1-\sigma}(t\sin(t\ln(2n))-(\sigma-1)\cos(t\ln(2n)))}{2\left(t^2+(\sigma-1)^2\right)}+C\right]_1^{\infty}$$

$$\int_1^{\infty}\cos(t\ln(2n))dn = \left[\frac{(2n)(t\sin(t\ln(2n))+\cos(t\ln(2n)))}{2(t^2+1)}+C\right]_1^{\infty}$$

For Gram[y=0] points-Dirichlet Sigma-Power Law, the equivalent of Eq. (9) and Eq. (11) are respectively given by Eq. (20) as equation and Eq. (21) as inequation.

$$-\frac{1}{2\left(t^2+(\sigma-1)^2\right)}\cdot[(2n)^{1-\sigma}((\sigma-1)\sin(t\ln(2n))+t\cos(t\ln(2n)))-$$

$$(2n-1)^{1-\sigma}((\sigma-1)\sin(t\ln(2n-1))+t\cos(t\ln(2n-1)))]_1^{\infty}=0 \quad (20)$$

$$\left[\frac{(2n)(\sin(t\ln(2n))-t\cos(t\ln(2n)))}{(2n-1)(\sin(t\ln(2n-1))-t\cos(t\ln(2n-1)))}-\frac{(2n)^{\sigma+1}}{(2n-1)^{\sigma+1}}\right]_1^{\infty} \neq$$

$$0$$

(21)

For Gram[x=0] points-Dirichlet Sigma-Power Law, the equivalent of Eq. (9) and Eq. (11) are respectively given by Eq. (22) as equation and Eq. (23) as inequation.

$$\frac{1}{2\left(t^2+(\sigma-1)^2\right)} \cdot [(2n)^{1-\sigma}\,(t\sin(t\ln(2n)) - (\sigma-1)\cos(t\ln(2n))) -$$

$$(2n-1)^{1-\sigma}\,(t\sin(t\ln(2n-1)) - (\sigma-1)\cos(t\ln(2n-1)))]_1^\infty = 0 \tag{22}$$

$$\left[\frac{(2n)(t\sin(t\ln(2n)) + \cos(t\ln(2n)))}{(2n-1)(t\sin(t\ln(2n-1)) + \cos(t\ln(2n-1)))} - \frac{(2n)^{\sigma+1}}{(2n-1)^{\sigma+1}}\right]_1^\infty \neq$$

$$0 \tag{23}$$

Intended derivation of Gram[y=0] and Gram[x=0] points-Dirichlet Sigma-Power Laws as equations and inequations is successful. *The proof is now complete for Lemma B.2*□.

Proposition B.3. Exact Dimensional analysis homogeneity at $\sigma = \frac{1}{2}$ in Gram[y=0] and Gram[x=0] points-Dirichlet Sigma-Power Laws as equations and inequations are (respectively) indicated by $\sum$(all fractional exponents) = whole number '1' and '3'.

Proof. Gram[y=0] points-Dirichlet Sigma-Power Law as equation for $\sigma = \frac{1}{2}$ value is given by:

$$-\frac{1}{2t^2+\frac{1}{2}} \cdot [(2n)^{\frac{1}{2}}\left(t\cos(t\ln(2n)) - \frac{1}{2}\sin(t\ln(2n))\right) -$$

$$(2n-1)^{\frac{1}{2}}\left(t\cos(t\ln(2n-1)) - \frac{1}{2}\sin(t\ln(2n-1))\right)]_1^\infty =$$

$$0 \tag{24}$$

Gram[y=0] points-Dirichlet Sigma-Power Law as inequation for $\sigma = \frac{1}{2}$ value is given by:

166

$$\left[\frac{(2n)\left(\sin\left(t\ln\left(2n\right)\right)-t\cos\left(t\ln\left(2n\right)\right)\right)}{(2n-1)\left(\sin\left(t\ln\left(2n-1\right)\right)-t\cos\left(t\ln\left(2n-1\right)\right)\right)}-\frac{(2n)^{\frac{3}{2}}}{(2n-1)^{\frac{3}{2}}}\right]_1^{\infty}\neq$$

$$0 \qquad\qquad (25)$$

Gram[x=0] points-Dirichlet Sigma-Power Law as equation for $\sigma = \frac{1}{2}$ value is given by:

$$\frac{1}{2t^2+\frac{1}{2})}\cdot[(2n)^{\frac{1}{2}}\left(t\,sin(t\,ln(2n))+\frac{1}{2}cos(t\,ln(2n))\right)$$

$$-(2n-1)^{\frac{1}{2}}(t\,sin(t\,ln(2n-1))+\frac{1}{2}cos(t\,ln(2n-1)))\}^{\infty}1_{\,=0}$$

$$(26)$$

Gram[x=0] points-Dirichlet Sigma-Power Law as inequation for $\sigma = \frac{1}{2}$ value is given by:

$$\left[\frac{(2n)\left(t\sin\left(t\ln\left(2n\right)\right)+\cos\left(t\ln\left(2n\right)\right)\right)}{(2n-1)\left(t\sin\left(t\ln\left(2n-1\right)\right)+\cos\left(t\ln\left(2n-1\right)\right)\right)}-\frac{(2n)^{\frac{3}{2}}}{(2n-1)^{\frac{3}{2}}}\right]_1^{\infty}\neq$$

$$0 \qquad\qquad (27)$$

$\sum$(all fractional exponents) as $2(1-\sigma)$ = whole number '1' for Eqs. (24) and (26), and $2(\sigma+1)$ = whole number '3' for Eqs. (25) and (27). These findings signify presence of complete sets Gram[y=0] points for Eqs. (24) and (25) and Gram[x=0] points for Eqs. (26) and (27). *The proof is now complete for Proposition B.3*□.

Corollary B.4. Inexact Dimensional analysis homogeneity at $\sigma = \frac{1}{2}$ [illustrated using $\sigma = \frac{2}{5}$] in Gram[y=0] and Gram[x=0] points-Dirichlet Sigma-Power Laws as equations and inequations are (respectively) indicated by $\sum$(all fractional exponents) = fractional number '≠ 1' and '≠ 3'.

Proof. Gram[y=0] points-Dirichlet Sigma-Power Law as equation for $\sigma = \frac{2}{5}$ value is given by:

$$-\frac{1}{2t^2 + \frac{18}{25}} \cdot \left[(2n)^{\frac{3}{5}}\left(t\cos(t\ln(2n)) - \frac{3}{5}\sin(t\ln(2n))\right) - \right.$$

$$\left.(2n-1)^{\frac{3}{5}}\left(t\cos(t\ln(2n-1)) - \frac{3}{5}\sin(t\ln(2n-1))\right)\right]_1^\infty =$$

0   (28)

Gram[y=0] points-Dirichlet Sigma-Power Law as inequation for $\sigma = \frac{2}{5}$ value is given by:

$$\left[\frac{(2n)(\sin(t\ln(2n)) - t\cos(t\ln(2n)))}{(2n-1)(\sin(t\ln(2n-1)) - t\cos(t\ln(2n-1)))} - \frac{(2n)^{\frac{7}{5}}}{(2n-1)^{\frac{7}{5}}}\right]_1^\infty \neq$$

0   (29)

Gram[x=0] points-Dirichlet Sigma-Power Law as equation for $\sigma = \frac{2}{5}$ value is given by:

$$\frac{1}{2t^2 + \frac{18}{25})} \cdot [(2n)^{\frac{3}{5}}\left(t\sin(t\ln(2n)) + \frac{3}{5}\cos(t\ln(2n))\right)$$

$$- (2n-1)^{\frac{3}{5}}\left(t\sin(t\ln(2n-1)) + \frac{3}{5}\cos(t\ln(2n-1))\right)]^\infty 1$$

$= 0$ (30)

Gram[x=0] points-Dirichlet Sigma-Power Law as inequation for $\sigma = \frac{2}{5}$ value is given by:

$$\left[\frac{(2n)(t\sin(t\ln(2n)) + \cos(t\ln(2n)))}{(2n-1)(t\sin(t\ln(2n-1)) + \cos(t\ln(2n-1)))} - \frac{(2n)^{\frac{7}{5}}}{(2n-1)^{\frac{7}{5}}}\right]_1^\infty \neq$$

0   (31)

$\sum$(all fractional exponents) as $2(1-\sigma)$ = fractional number '$\neq$ 1' for Eqs. (28) and (30), and $2(\sigma +1)$ = fractional number '$\neq$ 3' for Eqs. (29) and (31). These findings signify presence of complete sets virtual Gram[y=0] points for Eqs. (28) and (29) and virtual Gram[x=0] points for Eqs. (30) and (31). *The proof is now complete for Corollary B.4*□.

Appendix C: Hybrid method of Integer Sequence classification

The Hybrid method of Integer Sequence classification enables meaningful division of all integer sequences into either Hybrid or non-Hybrid integer sequences. My exotic A228186 [5] integer sequence was published on The On-line Encyclopedia of Integer Sequences website in 2013. It is the first ever [infinite length] Hybrid integer sequence synthesized from Combinatorics Ratio. In 'Position i' notation, let i = 0, 1, 2, 3, 4, 5,..., ∞ be complete set of natural numbers. A228186 "Greatest k > n such that ratio R < 2

is a maximum rational number with $R = \dfrac{Combinations\ With\ Repetition}{Combinations\ Without\ Repetition}$" is

equal to [infinite length] non-Hybrid (usual garden-variety) integer sequence A100967 except for finite 21 'exceptional' terms at Positions 0, 11, 13, 19, 21, 28, 30, 37, 39, 45, 50, 51, 52, 55, 57, 62, 66, 70, 73, 77, and 81 with their values given by relevant A100967 terms plus 1. The first 49 terms [from Position 0 to Position 48] of A100967 [6] "Least k such that binomial(2k+1, k-n) ≥ binomial(2k, k)" are listed below: 3, 9, 18, 29, 44, 61, 81, 104, 130, 159, 191, 225, 263, 303, 347, 393, 442, 494, 549, 606, 667, 730, 797, 866, 938, 1013, 1091, 1172, 1255, 1342, 1431, 1524, 1619, 1717, 1818, 1922, 2029, 2138, 2251, 2366, 2485, 2606, 2730, 2857, 2987, 3119, 3255, 3394, and 3535. For those 21 'exceptional' terms: at Position 0, A228186 (= 4) is given by A100967 (= 3) + 1; at Position 11, A228186 (= 226) is given by A100967 (= 225) + 1; at Position 13, A228186 (= 304) is given by A100967 (= 303) + 1; at Position 19, A228186 (= 607) is given by A100967 (= 606) + 1; etc. Here is a useful concept: Commencing from Position 0 onwards "in the limit" that this Position approaches 82, A228186 Hybrid integer sequence becomes (and is identical to) A100967 non-Hybrid integer sequence for all Positions ≥ 82.

169

Acknowledgements I am indebted to Mr. Rodney Williams and Mr. Tony O'Hagan for reviewing this paper (dedicated to my daughter Jelena born 13 weeks early on May 14, 2012).

References

1. Hardy, G. H. (1914), Sur les Zeros de la Fonction $\zeta(s)$ de Riemann, *C. R. Acad. Sci. Paris, 158:* 1012 – 1014, JFM 45.0716.04 Reprinted in (Borwein et al. 2008)

2. Hardy, G. H.; Littlewood, J. E. (1921), The zeros of Riemann's zeta-function on the critical line, *Math. Z., 10* (34): 283 – 317, http://dx.doi:10.1007/BF01211614

3. Abel, N.H. (1823), Solution de quelques problmes laide dintgrales dfinies, *Magazin Naturvidensk*, 1: 55 – 68

4. Plana, G.A.A. (1820), 'Sur une nouvelle expression analytique des nombres Bernoulliens, propre exprimer en termes finis la formule gnrale pour la sommation des suites', *Mem. Accad. Sci. Torino*, 25: 403 – 418

5. Ting, J (August 15, 2013), A228186, OEIS Foundation Inc. (2011), The On-Line Encyclopedia of Integer Sequences, https://oeis.org/A228186

6. Noe, T (November 23, 2004), A100967, OEIS Foundation Inc. (2011), The On-Line Encyclopedia of Integer Sequences, https://oeis.org/A100967

Appendix 2 Solving Incompletely Predictable problems Polignac's and Twin Prime conjectures using Information-Complexity

Solving Incompletely Predictable problems Polignac's and Twin Prime conjectures using Information-Complexity conservation

John Y. C. Ting

Published in viXra on April 26, 2019

Abstract Prime numbers are Incompletely Predictable numbers calculated using complex algorithm Sieve of Eratosthenes. Involving proposals that prime gaps and associated sets of prime numbers are infinite in magnitude, Twin prime conjecture deals with even prime gap 2 and is a subset of Polignac's conjecture which deals with all even prime gaps 2, 4, 6, 8, 10,.... Treated as Incompletely Predictable problems, we solve these conjectures with research method Information-Complexity conservation to get Plus Gap 2 Composite Number Continuous Law and Plus-Minus Gap 2 Composite Number Alternating Law.

Keywords Dimensional analysis, Incompletely Predictable problems, Information-Complexity conservation, Plus Gap 2 Composite Number Continuous Law, Plus-Minus Gap 2 Composite Number Alternating Law, Polignac's and Twin prime conjectures

2010 Mathematics Subject Classification. 11A41, 11M26

1 Introduction

Uncountable complex numbers (C) include uncountable real numbers (R). R = countable rational numbers (Q) + uncountable irrational numbers (R – Q). R – Q = countable algebraic numbers + uncountable transcendental numbers. Q include countable integers (Z) which include countable whole numbers (W) which in turn include countable natural numbers (N). N is constituted by either countable even numbers (E) and countable odd numbers (O) or countable prime numbers (P), countable composite numbers (C) and Number '1'. Then (i) Set N = Set E + Set O, (ii) Set N = Set P + Set C + Number '1', and (iii) Set N ⊂ Set W ⊂ Set Z ⊂ Set Q ⊂ Set R ⊂ Set C.

With increasing magnitude, arbitrary Set X belongs to countable finite set (CFS), countable infinite set (CIS) or uncountable infinite set (UIS). Cardinality of Set X, $|X|$, measures the "number of elements" in Set X. E.g. Set even P has CFS of even P with $|even\ P| = 1$, Set N has CIS of N with $|N| = \aleph_0$, and Set R has UIS of R with $|R| =$ c (cardinality of the continuum). Respectively, CIS of P and C are *Incompletely Predictable numbers* dependently calculated directly and indirectly from *complex algorithm* Sieve of Eratosthenes. Involving proposals that prime gaps and associated sets of prime numbers are infinite in magnitude, Twin prime conjecture deals with even prime gap 2 and is a subset of Polignac's conjecture dealing with all even prime gaps 2, 4, 6, 8, 10,.... Activities to prove these open problems in number theory equate to solving *Incompletely Predictable problems*.

All claims arising from these activities are made meaningful by providing definitions on above mentioned terms. Respectively, an Incompletely (Completely) Predictable number is locationally defined as a number whose position is *dependently* (*independently*) determined by complex (simple) calculations using complex (simple) equation or algorithm with (without) needing to know related positions of all preceding numbers in neighborhood. Simple properties are inferred from a phrase such as: "...simple equation or algorithm by itself will intrinsically incorporate actual location [and actual positions] of all Completely Predictable numbers". Solving Completely Predictable problems with simple properties amendable to *simple* treatments using *usual* mathematical tools such as Calculus result in 'Simple Elementary Fundamental Laws'-based solutions. Complex properties are inferred from a phrase such as: "...complex equation or algorithm by itself will intrinsically incorporate actual location [but not actual positions] of all Incompletely Predictable numbers". Solving Incompletely Predictable problems with complex properties amendable to *complex* treatments using *unusual* mathematical tools such as our novel research method Information-Complexity conservation as well as using *usual* mathematical tools such as Calculus result in 'Complex Elementary Fundamental Laws'-based solutions.

1.1 Dimensional analysis on Cardinality and "Dimensions"

For 'base quantities' such as *length, mass* and *time*, their fundamental SI 'units of measurement' are [respectively] given by meter (m), kilogram (kg) and second (s). The word 'dimension' is commonly used to denote 'units of measurement' in well-defined equations. Dimensional analysis (DA) is an analytic tool with resulting DA homogeneity and nonhomogeneity (respectively) denoting valid and invalid equation when 'units of measurements' are "balanced" and "unbalanced" across both sides of the equation. E.g. 2 m + 3 m = 5 m is a valid equation but 2 m + 3 kg = 5 mkg is an invalid equation.

We use "Dimensions" to denote well-defined Incompletely Predictable entities obtained from Information-Complexity conservation. Relevant "Dimensions" *dependently* represent Number '1', P and C. Then *by default* any (sub)sets of P and C in well-defined equations can also be represented by their corresponding "Dimensions".

Remark 1.1. We can apply Dimensional analysis to "Dimensions" from Information-Complexity conservation and cardinality of relevant sets in certain well-defined equations.

Let X denote E, O, N [which are classified as Completely Predictable numbers], P and C [which are classified as Incompletely Predictable numbers]. For x = 1, 2, 3, 4, 5,..., ∞; consider all X $\leq$ x. Then this "all X $\leq$ x" is definition for X-$\pi(x)$ [denoting "X counting function"] resulting in following two types of equations coined as (I) 'Exact' equation N$\pi(x)$ = E-$\pi(x)$ + O-$\pi(x)$ with "non-varying" relationships E-$\pi(x)$ = O-$\pi(x)$ for all x = E and E-$\pi(x)$ = O-$\pi(x)$ - 1 for all x = O, and (II) 'Inexact' equation N-$\pi(x)$ = 1 + P-$\pi(x)$ + C-$\pi(x)$ with "varying" relationships P-$\pi(x)$ > C-$\pi(x)$ for all x $\leq$ 8; P-$\pi(x)$ = C-$\pi(x)$ for x = 9, 11, and 13; and P-$\pi(x)$ < C-$\pi(x)$ for x = 10, 12, and all x $\geq$ 14.

173

Let "Dimensions" and different (sub)sets of E, O, N, P and C be 'base quantities'. Then exponent '1' of "Dimensions" and cardinality of these (sub)sets in well-defined equations are corresponding 'units of measurement'. Performing DA on "Dimensions" for PC pairing are depicted later on. Performing DA on cardinality are depicted next.

For Set N = Set E + Set O, then $|N| = |E| + |O| \Longrightarrow \aleph_0 = \aleph_0 + \aleph_0$ thus conforming with DA homogeneity.

For Set N = Set P + Set C + Number '1', then Set N - Number '1' = Set P + Set C and $|N - \text{Number '1'}| = |P| + |C| \Longrightarrow \aleph_0 = \aleph_0 + \aleph_0$ thus conforming with DA homogeneity.

For Set N - Set even P - Number '1'= Set odd P + Set even C + Set odd C, then $|N - \text{even P} - \text{Number '1'}| = |\text{odd P}| + |\text{even C}| + |\text{odd C}| \Longrightarrow \aleph_0 = \aleph_0 + \aleph_0 + \aleph_0$ thus conforming with DA homogeneity. Symbolically represented by all available O prime gap = 1 and E prime gaps = 2, 4, 6, 8, 10,...; O composite gap = 1 and E composite gap = 2; and O natural gap = 1; then $|\text{Gap 1 N} - \text{Gap 1 P} - \text{Number '1'}| = |\text{Gap 2 P}| + |\text{Gap 4 P}| + |\text{Gap 6 P}| + |\text{Gap 8 P}| + |\text{Gap 10 P}| + ... + |\text{Gap 1 C}| + |\text{Gap 2 C}| \Longrightarrow \aleph_0 = \aleph_0 + \aleph_0 + \aleph_0 + \aleph_0 + \aleph_0 + ... \aleph_0 + \aleph_0$ thus conforming with DA homogeneity. It is known that $|\text{Gap 1 P}| = |\text{Number '1'}| = 1$ and $|\text{Gap 1 N}| = |\text{Gap 1 C}| = |\text{Gap 2 C}| = \aleph_0$. Then solving Polignac's and Twin prime conjectures translate to successfully proving $|\text{Gap 2 P}| = |\text{Gap 4 P}| = |\text{Gap 6 P}| = |\text{Gap 8 P}| = |\text{Gap 10 P}| = ... = \aleph_0$ with $|\text{E prime gaps}| = \aleph_0$.

Outline of proof for Polignac's and Twin prime conjectures. Requires simultaneously satisfying two mutually inclusive conditions:

I. With rigid manifestation of DA homogeneity, quantitive[1] fulfillment by considering i ∈ E for each Subset odd Pi generated by E prime gap = i from Set E prime gaps occurs only if solitary cardinality value is present in equation Set odd P =

$$\sum_{i=2}^{\infty} Subset\ odd\ Pi$$

with $|\text{odd P}| = |\text{odd Pi}| = |\text{E prime gaps}| = \aleph_0$.

II. With rigid manifestation of DA non-homogeneity, quantitive[1] fulfillment by considering i ∈ E for each Subset odd Pi generated by E prime gap = i from Set E prime gaps does not occur if more than one cardinality values are present in equation

$$\text{Set odd P} > \sum_{i=2}^{\infty} \textit{Subset odd Pi}$$

with |E prime gaps| = $\aleph_0$ having incorrect

$$\text{Set odd P} > \sum_{i=2}^{N} \textit{Subset odd Pi}$$

|Subset(s) odd P| = N (finite value) and/or with |odd Pi| = $\aleph_0$ having incorrect |E prime gaps| = N (finite value).

Footnote 1: Qualitative fulfilment of |odd P| = |odd P_i| = |all E prime gaps| = $\aleph_0$ equates to Plus-Minus Gap 2 Composite Number Alternating Law being precisely obeyed by all E prime gaps apart from first E prime gap precisely obeying Plus Gap 2 Composite Number Continuous Law. Derived using Information-Complexity conservation, these Laws symbolize "end-result" proof on Polignac's and Twin prime conjectures. *Law of Continuity* is a heuristic principle *whatever succeed for the finite, also succeed for the infinite.* Then these Laws which inherently manifest 'Gap 2 Composite Number' on finite and infinite time scale should in principle "succeed for the finite, also succeed for the infinite".

Polignac's and Twin prime conjectures mathematical foot-prints. Six identifiable steps to prove these conjectures: *Step 1* Considering x ∈ N, obtain Dimensions $(2x - 2)^1$, $(2x - 4)^1$, $(2x - 5)^1$, $(2x - 7)^1$, $(2x - 8)^1$, $(2x - 9)^1$, ..., $(2x - \infty)^1$ with specific groupings to constitute all elements of Set P [culminating in obtaining all prime gaps (= E prime gaps + Solitary O prime gap) with |all prime gaps| = $\aleph_0$]. Note Dimension $(2x - 2)^1$ represents x = 1 (Number '1') which is neither P nor C. *Step 2* Considering i ∈ E, confirm perpetual recurrences of individual E prime gap = i (associated with its unique odd P_i) occur only when depicted as specific groupings of these Dimensions endowed with exponent '1' for all ranges of x. *Step 3* Perform DA on exponent '1' in these Dimensions. *Step 4* Perform DA on equation

Set odd P $= \sum\limits^{\infty}$Subset odd P_i to obtain $|\textbf{odd P}| = |\textbf{odd P}_i| = \aleph_0$ whereby Subset odd P_i is derived from its associated unique E prime gap $= i$ with $|\textbf{E prime gaps}| = \aleph_0$. *Step 5* Confirm 'Prime number' variable and 'Prime gap' variable complex algorithm "containing" all P with knowing their overall actual location [but not actual positions][2]. *Step 6* Derive Plus-Minus Gap 2 Composite Number Alternating Law and Plus Gap 2 Composite Number Continuous Law using Information-Complexity conservation.

Footnote 2: This phrase implies all P (and C) are treated as Incompletely Predictable numbers. Actual positions will require using complex algorithm Sieve of Eratosthenes to *dependently* calculate positions of all preceding P (and C) in the neighborhood.

'Complex Elementary Fundamental Laws'-based solutions of Plus-Minus Gap 2 Composite Number Alternating Law and Plus Gap 2 Composite Number Continuous Law are obtained by undertaking the non-negotiable mathematical steps outlined above. These Laws are literally Completely Predictable meta-properties ('overall' *complex properties*) arising from "interactions" between P and C producing relevant patterns of Gap 2 Composite Number perpetual appearances [albeit with Incompletely Predictable timing]. We logically deduce explicit mathematical explanation for these meta-properties requires "complex" mathematical arguments. Attempts to give explicit mathematical explanation with "simple" mathematical arguments would intuitively mean Incompletely Predictable numbers P and C be (incorrectly and impossibly) treated as Completely Predictable numbers.

1.2 Brief overview of Polignac's and Twin prime conjectures

Occurring over 2000 years ago (c. 300 BC), ancient Euclid's proof on infinitude of P in totality [viz. $|P| = \aleph_0$ for Set P] predominantly by *reductio ad absurdum* (proof by contradiction) is earliest known but not the only proof for this simple problem in number theory. Since then dozens of proofs have been devised such as three chronologically listed: Goldbach's Proof using Fermat numbers (written in a letter to Swiss mathematician Leonhard Euler, July 1730), Furstenberg's Topological Proof in

1955[1], and Filip Saidak's Proof in 2006[2]. The strangest candidate is likely to be Furstenberg's Topological Proof.

In 2013, Yitang Zhang proved a landmark result showing some unknown even number 'N' < 70 million such that there are infinitely many pairs of P that differ by 'N'[3]. By optimizing Zhang's bound, subsequent Polymath Project collaborative efforts using a new refinement of GPY sieve in 2013 lowered 'N' to 246; and assuming Elliott-Halberstam conjecture and its generalized form have further lower 'N' to 12 and 6, respectively. Then 'N' has intuitively more than one valid values such that there are infinitely many pairs of P that differ by each of those 'N' values [thus proving existence of more than one Subset **odd P_i** with $|\text{\textbf{odd }} \mathbf{P}_i| = \aleph_0$]. We can only theoretically lower 'N' to 2 (in regards to P with 'small gaps') but there are still an infinite number of E prime gaps (in regards to P with 'large gaps') that require "the proof that each will generate its unique set of infinite P".

Remark 1.2. Existence of maximal and non-maximal prime gaps supply crucial indirect evidence to intuitively support but does not prove proposition "Each even prime gap will generate an infinite magnitude of odd prime numbers on its own accord". Comments relevant to Remark 1.2 are given in Section 2 below.

2 Supportive role of maximal and non-maximal prime gaps

We analyze data of all P obtained when extrapolated out over a wide range of x ≥ 2 integer values. As sequence of P carries on, P with ever larger prime gaps will appear. For given range of x integer values, prime gap = n2 is a 'maximal prime gap' if prime gap = n1 < prime gap = n2 for all n1 < n2. In other words, the largest such prime gaps in this range are called maximal prime gaps. The term 'first occurrence prime gaps' refers to first occurrences of maximal prime gaps whereby maximal prime gaps are prime gaps of "at least of this length".

177

Table 1 First 17 prime gaps depicted in the format utilizing maximal prime gaps [depicted with asterisk symbol (*)] and non-maximal prime gaps [depicted without this asterisk symbol].

Prime gap	Following the prime number	Prime gap	Following the prime number
1*	2	18*	523
2*	3	20*	887
4*	7	22*	1129
6*	23	24	1669
8*	89	26	2477
10	139	28	2971
12	199	30	4297
14*	113	32	5591
16	1831		

We use maximal prime gaps to denote 'first occurrence prime gaps'. CIS non-maximal prime gaps (endorsed with nickname 'slow jumpers') will always lag behind CIS maximal prime gaps for onset appearances in P sequence. These are shown for first 17 prime gaps in Table 1. Apart from O prime gap = 1 representing solitary even P '2', remaining P depicted in Table 1 consist of representative single odd P for each E prime gap. These odd P will individually make one-off appearance in P sequence in a *perpetual albeit Incompletely Predictable manner*. Initial seven of [majority] "missing" odd P are 5, 11, 13, 17, 19, 29, 31,... belonging to Subset P with 'residual' prime gaps are potential source of odd P in relation to proposal that each E prime gap from Set E prime gaps will generate its specific Subset odd P. Set all P from all prime gaps = Subset P from maximal prime gaps + Subset P from non-maximal prime gaps + Subset P from 'residual' prime gaps. Subset P from 'residual' prime gaps with representation from all E prime gaps must include all correctly selected "missing" odd P. These observations support but does not prove proposition that each E prime gap will generate its own Subset odd P with |odd P| = $\aleph_0$.

For i ∈ N; primordial $P_i\#$ is analog of usual factorial for P = 2, 3, 5, 7, 11, 13,.... Then $P_1\# = 2$, $P_2\# = 2 \times 3 = 6$, $P_3\# = 2 \times 3 \times 5 = 30$, $P_4\# = 2 \times 3 \times 5 \times 7 = 210$, $P_5\# = 2 \times 3 \times 5 \times 7 \times 11 = 2310$, $P_6\# = 2 \times 3 \times 5 \times 7 \times 11 \times 13 = 30030$, etc. English mathematician John Horton Conway coined the term 'jumping champion' in 1993. An integer n is a 'jumping champion' if n is the most frequently occurring difference (prime gap) between consecutive P<x for some x integer values. Example: for any x with 7<x<131, n = 2 (indicating twin P) is the 'jumping champion'. It has been conjectured that (i) the only 'jumping champions' are 1, 4 and primorials 2, 6, 30, 210, 2310, 30030,... and (ii) 'jumping champions' tend to infinity. Their required proofs will likely need proof of k-tuple conjecture. P from 'jumping champion' prime gaps have their onset appearances in P sequence in a *perpetual albeit Incompletely Predictable manner* [as another example to that outlined in previous paragraph].

3 Information-Complexity conservation

A formula, as equation or algorithm, is simply a Black Box generating necessary Output (with qualitative structural 'Complexity') when supplied with given Input (with quantitative data 'Information'). This 'Information' and 'Complexity' are what is referred to in the term 'Information-Complexity conservation'.

N (CIS): 1, 2, 3,..., +∞. Let x be from Set X such that x ∈ N. Consider x for the upper boundary of interest in Set X whereby X is chosen from N, E, O, P or C.

Lemma 3.1. Natural counting function N-$\pi(x)$, defined as |N ≤ x|, is Completely Predictable by independently using simple algorithm to be equal to x.

Proof Formula to generate N with 100% certainty is $N_i = i$ whereby N_i is the ith N and i = 1, 2, 3,..., ∞. For a given N_i, its ith position is simply i. Natural gap $(G_{Ni}) = N_{i+1} - N_i$, with G_{Ni} always = 1. There are x N ≤ x. Thus N-$\pi(x)$ = |N ≤ x| = x. *The proof is now complete for Lemma 3.1*□.

179

Lemma 3.2. Even counting function E-$\pi(x)$, defined as $|E \leq x|$, is Completely Predictable by independently using simple algorithm to be equal to floor(x/2).

Proof. Formula to generate E with 100% certainty is $E_i = i \times 2$ whereby E_i is the ith E and i = 1, 2, 3,..., ∞ abiding to mathematical label "All N always ending with a digit 0, 2, 4, 6 or 8". For a given E_i, its ith position is calculated as i = $E_i/2$. Even gap (G_{Ei}) = E_{i+1} - E_i, with G_{Ei} always = 2. There are $\lfloor \frac{x}{2} \rfloor$ E $\leq$ x. Thus E-$\pi(x)$ = $|E \leq x|$ = floor(x/2). *The proof is now complete for Lemma 3.2*□.

Lemma 3.3. Odd counting function O-$\pi(x)$, defined as $|O \leq x|$, is Completely Predictable by independently using simple algorithm to be equal to ceiling(x/2).

Proof. Formula to generate O with 100% certainty is O_i = (i $\times$ 2) - 1 whereby O_i is the ith odd number and i = 1, 2, 3,..., ∞ abiding to mathematical label "All N always ending with a digit 1, 3, 5, 7, or 9". For a given O_i number, its ith position is calculated as i = $(O_i + 1)/2$. Odd gap (G_{Oi}) = O_{i+1} - O_i, with G_{Oi} always = 2. There are $\lceil \frac{x}{2} \rceil$ O $\leq$ x. Thus O-$\pi(x)$ = $|O \leq x|$ = ceiling(x/2). *The proof is now complete for Lemma 3.3*□.

Lemma 3.4. Prime counting function P-$\pi(x)$, defined as $|P \leq x|$, is Incompletely Predictable with Set P dependently obtained using complex algorithm Sieve of Eratosthenes.

Proof. Algorithm to generate P_i whereby P_1 (= 2), P_2 (= 3), P_3 (= 5), P_4 (= 7),..., ∞ with 100% certainty is based on Sieve of Eratosthenes abiding to mathematical label "All N apart from 1 that are evenly divisible by itself and by 1". Although we can check primality of a given O by trial division, we can never determine its position without knowing positions of preceding P. Prime gap (G_{Pi}) = P_{i+1} - P_i, with G_{Pi} constituted by all E except 1st G_{P1} = 3 - 2 = 1. P-$\pi(x)$ = $|P \leq x|$. This is Incompletely Predictable and is calculated via mentioned algorithm. Using definition of prime gap, every P [represented here with aid of 'n' notation instead of usual 'i' notation] is written as

$$P_{n+1} = 2 + \sum_{i=1}^{n} GPi$$ with '2' denoting P_1. Here i & n = 1, 2, 3, 4, 5, ..., ∞. *The proof is now complete for Lemma 3.4□.*

Lemma 3.5. Composite counting function $C\text{-}\pi(x)$, defined as $|C \leq x|$, is Incompletely Predictable with Set C derived as Set N - Set P [dependently obtained using complex algorithm Sieve of Eratosthenes] - Number '1'.

Proof. Composite numbers abide to mathematical label "All N apart from 1 that are evenly divisible by numbers other than itself and 1". Algorithm to generate C_i whereby C_1 (= 4), C_2 (= 6), C_3 (= 8), C_4 (= 9),..., ∞ with 100% certainty is based [indirectly] on Sieve of Eratosthenes via selecting non-prime N to be C. We define Composite gap G_{Ci} as $C_{i+1} - C_i$ with G_{Ci} constituted by 1 & 2. $C\text{-}\pi(x) = C \leq x$. This is Incompletely Predictable and always need to be calculated indirectly via mentioned algorithm. Using definition of composite gap, every C [represented here with aid of 'n' notation instead usual 'i' notation] is written as $C_{n+1} = 4 + \sum_{i=1}^{n} GCi$ with '4' denoting C_1. Here i & n = 1, 2, 3, 4, 5, ..., ∞. *The proof is now complete for Lemma 3.5□.*

Denote X to be N, E, O, P or C. $X\text{-}\pi(x) = |X \leq x|$ with x ∈ N. We define and compute entity 'Grand-Total Gaps for X at x' (Grand-Total ΣX_x-Gaps).

Proposition 3.6. For any given x ≥ 1 values in Set N, designated Complexity is represented by ΣN_x-Gaps = x - N with N = 1 being maximal.

Proof. Set N (for x = 1 to 12): 1, 2, 3, 4, 5, 6, 7, 8, 9, 10, 11, 12. $N\text{-}\pi(x) = 12$. There are x - 1 = 11 N-Gaps each of '1' magnitude: 1, 1, 1, 1, 1, 1, 1, 1, 1, 1, 1. ΣN_x-Gaps = 11 X 1 = 11. This equates to "x - 1" – regarded as Complexity for N. *The proof is now complete for Proposition 3.6□.*

Proposition 3.7. For any given x ≥ 1 values in constituent Set E and Set O, designated Complexity is represented by ΣEO_x-Gaps = 2x - N with N = 4 being maximal.

Proof. Set E and Set O (for x = 1 to 12): 2, 4, 6, 8, 10, 12 and 1, 3, 5, 7, 9, 11. E-$\pi(x)$

= 6 and O-$\pi(x)$ = 6. There are $\lfloor \frac{x}{2} \rfloor$ - 1 = 5 E-Gaps each of '2' magnitude: 2, 2, 2, 2, 2.

ΣE_x-Gaps = 5 X 2 = 10, and $\lceil \frac{x}{2} \rceil$ - 1 = 5)-Gaps each of '2' magnitude: 2, 2, 2, 2, 2. ΣO_x-Gaps = 5 X 2 = 10. Grand-Total ΣEO_x-Gaps = 10 + 10 = 20. Depicted by Table 3 and Figure 2 in Appendix I, 2x - N = "2x - 4" [perpetual constant appearances of "N = 4 being maximal"] is Complexity for E and O. *The proof is now complete for Proposition 3.7*□.

Proposition 3.8. For selected x ≥ 2 values in constituent Set P and Set C, designated Complexity is cyclically represented by ΣPC_x-Gaps = 2x - N with N = 7 being minimal.

Proof. Set P and Set C (for x = 2 to 12): 2, 3, 5, 7, 11 and 4, 6, 8, 9, 10, 12. P-$\pi(x)$ = 5 and C-$\pi(x)$ = 6. There are four P-Gaps of 1, 2, 2, 4 magnitude and five C-Gaps of 2, 2, 1, 1, 2 magnitude. ΣP_x-Gaps = 1 + 2 + 2 + 4 = 9. ΣC_x-Gaps = 2 + 2 + 1 + 1 + 2 = 8. Grand-Total ΣPC_x-Gaps = 9 + 8 = 17. Depicted by Table 2 and Figure 1, 2x - N = 2x - 7 [perpetual intermittent and cyclical appearances of "N = 7 being minimal"] is Complexity for P and C. *The proof is now complete for Proposition 3.8*□.

Designated Complexity is (i) x - N with N = 1 (maximal) for Completely Predictable N, (ii) 2x - N with N = 7 (minimal) for Incompletely Predictable P & C, and (iii) 2x - N with N = 4 (maximal) for Completely Predictable E & O. Interpretations: N has minimal Complexity, E & O have intermediate Complexity, and P & C have maximal [varying] Complexity. Defacto baseline "2x - 4" Grand-Total Gaps [minus 4 value] in E-O pairing > Defacto baseline "2x - ≥7" Grand-Total Gaps [minus ≥7 values] in P-C pairing.

Let both x & N ∈ N. We tabulate in Table 2 and graph in Figure 1 [Incompletely Predictable] P-C mathematical landscape for a relatively larger x = 2 to 64 here (and ditto for [Completely Predictable] E-O mathematical landscape for relatively larger x = 1 to 64 in Appendix I). The term "mathematical landscape" denotes specific mathematical patterns in tabulated and graphed data. "Dimension" contextually denotes Dimension 2x - N whereby (i) allocated [infinite] N values result in Dimensions 2x - 7, 2x - 8, 2x - 9, ..., 2x - ∞ for P-C finite scale mathematical landscape and (ii) allocated [finite] N values for E-O finite scale mathematical landscape result in Dimension 2x - 4. For P-C pairing, initial one-off Dimensions 2x - 2, 2x - 4 and 2x - 5 (in consecutive order) are exceptions [with Dimension 2x - 2 validly representing Number '1' which is neither P nor C]. For E-O pairing, initial one-off Dimension 2x - 2 is an exception. P-C mathematical landscape consisting of Dimensions will intrinsically incorporate P and C in an integrated manner and there are infinite times whereby relevant Dimensions deviate away from 'baseline' Dimension 2x - 7 simply because P [and, by default, C] in totality are rigorously proven to be infinite in magnitude. In contrast, there is complete lack of deviation away from 'baseline' Dimension 2x - 4 apart from one-off deviation by initial Dimension 2x - 2 in Appendix I.

Table 2 Prime-Composite finite scale mathematical (tabulated) landscape using data obtained for x = 2 to 64. The Number '1' is neither prime nor composite. Legend: C = composite, P = prime, Y = Dimension 2x 7 (for visual clarity), N/A = Not Applicable.

x	P_i or C_i, Gaps	ΣPC_{x^-} Gaps	Dimension
1	N/A	0	2x-2
2	P1, 1	0	2x-4
3	P2, 2	1	2x-5
4	C1, 2	1	Y
5	P3, 2	3	Y
6	C2, 2	5	Y
7	P4, 4	7	Y
8	C3, 1	9	Y
9	C4, 1	10	2x-8
10	C5, 2	11	2x-9

11	P5, 2	15	Y
12	C6, 2	17	Y
13	P6, 4	19	Y
14	C7, 1	21	Y
15	C8, 1	22	2x-8
16	C9, 1	23	2x-9
17	P7, 2	27	Y
18	C10, 2	29	Y
19	P8, 4	31	Y
20	C11, 1	33	Y
21	C12, 1	34	2x-8
22	C13, 2	35	2x-9
23	P9, 6	39	Y
24	C14, 1	41	Y
25	C15, 1	42	2x-8
26	C16, 1	43	2x-9
27	C17, 1	44	2x-10
28	C18, 2	45	2x-11
29	P10, 2	51	Y
30	C19, 2	53	Y
31	P11, 6	55	Y
32	C20, 1	57	Y
33	C21, 1	58	2x-8
34	C22, 1	59	2x-9
35	C23, 1	60	2x-10
36	C24, 2	61	2x-11
37	P12, 4	67	Y

38	C25, 1	69	Y
39	C26, 1	70	2x-8
40	C27, 1	71	2x-9
41	P13, 2	75	Y
42	C28, 2	77	Y
43	P14, 4	79	Y
44	C29, 1	81	Y
45	C30, 1	82	2x-8
46	C31, 2	83	2x-9
47	P15, 6	87	Y
48	C32, 1	89	Y
49	C33, 1	90	2x-8
50	C34, 1	91	2x-9
51	C35, 1	92	2x-10
52	C36, 1	93	2x-11
53	P16, 6	99	Y
54	C37, 1	101	Y
55	C38, 1	102	2x-8
56	C39, 1	103	2x-9
57	C40, 1	104	2x-10
58	C41, 1	105	2x-11
59	P17, 2	111	Y
60	C42, 2	113	Y
61	P18, 6	115	Y
62	C43, 1	117	Y
63	C44, 1	118	2x-8
64	C45, 1	119	2x-9

In Figure 1, Dimensions 2x - 7, 2x - 8, 2x - 9, ..., 2x - ∞ are symbolically represented by -7, -8, -9, ..., ∞ with 2x - 7 displayed as 'baseline' Dimension whereby Dimension trend (Cumulative Sum Gaps) must repeatedly reset itself onto this 'baseline' Dimension on a perpetual basis. Dimensions symbolically represented by ever larger negative integers will correspond to P associated with ever larger prime gaps and this phenomenon will generally happen at ever larger x values (with complete presence of Chaos and Fractals being manifested in our graph). At ever larger x values, P-$\pi(x)$ will overall become larger but with a *decelerating* trend whereas C-$\pi(x)$ will overall become larger but with an *accelerating* trend. This support ever larger prime gaps appearing at ever larger x values.

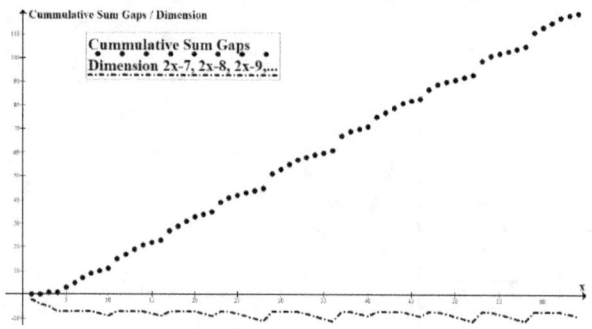

Fig. 1 Prime-Composite finite scale mathematical (graphed) landscape using data obtained for x = 2 to 64. Bottom graph symbolically represent "Dimensions" using ever larger negative integers.

Definitive derivation of data in Table 2 is illustrated by two examples for position x = 31 & 32. For i & x ∈ N; ΣPC_x-Gap = ΣPC_{x-1}-Gap + Gap value at P_{i-1} or Gap value at C_{i-1} whereby (i) P_i or C_i at position x is determined by whether relevant x value belongs to a P or C, and (ii) both ΣPC_1-Gap and ΣPC_2-Gap = 0. Example, for position x = 31: 31 is P (P11). Desired Gap value at P10 = 2. Thus ΣPC_{31}-Gap (55) = ΣPC_{30}-Gap (53) + Gap value at P10 (2). Example, for position x = 32: 32 is C (C20). Desired Gap value at C19 = 2. Thus ΣPC_{32}-Gap (57) = ΣPC_{31}-Gap (55) + Gap value at C20 (2). 'Overall magnitude of C will always be greater than that of P' will hold true from x = 14 onwards. For instance, position x = 61 corresponds to P 61 which is 18th P, whereas [the one lower] position x = 60 corresponding to C 60 is the [much higher] 42nd C.

4 Polignac's and Twin prime conjectures

Previous section alludes to P-C finite scale mathematical landscape. This section alludes to P-C infinite scale mathematical landscape. Let 'Y' symbolizes (baseline) Dimension 2x - 7. Let prime gap at $P_i = P_{i+1} - P_i$ with P_i & P_{i+1} respectively symbolizes consecutive "first" & "second" P in any P_i-P_{i+1} pairings. We denote (i) Dimensions YY grouping [depicted by 2x - 7 initially appearing twice in (iii)] to represent signal for appearances of P pairings other than twin P such as cousin P, sexy P, etc; (ii) Dimension YYYY grouping to represent signal for appearances of P pairings as twin P; and (iii) Dimension (2x - ≥7)-Progressive Grouping allocated to 2x - 7, 2x - 7, 2x - 8, 2x - 9, 2x - 10, 2x - 11,..., 2x - ∞ as elements of *precise* and *proportionate* CFS Dimensions representation of an individual P_i with its associated prime gap namely, Dimensions 2x - 7 & 2x - 7 pairing = twin P (with both its prime gap & CFS cardinality = 2); 2x - 7, 2x - 7, 2x - 8 & 2x - 9 pairing = cousin P (with both its prime gap & CFS cardinality = 4); 2x - 7, 2x - 7, 2x - 8, 2x - 9, 2x - 10 & 2x - 11 pairing = sexy P (with both its prime gap & CFS cardinality = 6); and so on. The higher order [traditionally defined as closest possible] prime groupings of three P as P triplets, of four P numbers as prime quadruplets, of five P numbers as prime quintuplets, etc consist of relevant serendipitous groupings abiding to mathematical rule: With exception of three 'outlier' P 3, 5, & 7; groupings of any three P as P, P+2, P+4 combination (viz. manifesting two consecutive twin P) is a mathematical impossibility. The 'anomaly' one of every three consecutive O is a multiple of three, and hence this particular number cannot be P, explains this impossibility. Then closest possible P grouping [viz. for prime triplet] must be either P, P+2, P+6 format or P, P+4, P+6 format.

P groupings not respecting traditional closest-possible-prime groupings are also the norm occurring infinitely often, indicating continual presence of prime gaps ≥ 6. As P become sparser at larger range, perpetual presence of (i) prime gaps ≥ 6 [which we

187

propose to arbitrarily represent 'large gaps'] and (ii) prime gaps 2 & 4 [which we propose to arbitrarily represent 'small gaps'] with progressive greater magnitude will cumulatively occur for each prime gap but always in a decelerating manner. With permanent requirement at larger range of intermittently resetting to baseline Dimension 2x - 7 occurring [either two or] four times in a row, nature seems to dictate, at the very least, perpetual twin P or one other non-twin P occurrences is inevitable.

We dissect Dimension YYYY unique signal for twin P appearances. Initial two CFS Dimensions YY components of YYYY represent "first" P component of twin P pairing. Last two Dimensions YY components of YYYY signifying appearance of "second" P component of twin P pairing is also the initial first-two-element component of full CFS Dimensions representation for "first" P component of following non-twin P pairing. Twin P are uniquely represented by repeating *single* type Dimension 2x - 7. In all other 'higher order' P pairings (with prime gaps $\geq$ 4), they require *multiple* types Dimension representation. There is qualitative aspect association of *single* type Dimension representation for twin P resulting in "less colorful" Plus Gap 2 Composite Number *Continuous Law* as opposed to *multiple* types Dimension representation for all other 'higher order' P pairings resulting in "more colorful" Plus-Minus Gap 2 Composite Number *Alternating Law*. 'Gap 2 Composite Number' occurrences in both Laws on finite scale are (directly) observed in Figure 1 & Table 2 for x = 2 to 64, and on infinite scale are (indirectly) deduced using logical arguments for all x values.

We endow all "Dimensions" with exponent of '1' for perusal in on-going mathematical arguments. P_1 = 2 is represented by CFS as Dimension $(2x - 4)^1$ (with both prime gap & CFS cardinality = 1); P_2 = 3 is represented by CFS as Dimensions $(2x - 5)^1$ & $(2x - 7)^1$ (with both prime gap & CFS cardinality = 2); P_3 = 5 is represented by CFS Dimension $(2x - 7)^1$ & $(2x - 7)^1$ (with both prime gap & CFS cardinality = 2), etc.

Proposition 4.1. Let Case 1 be Completely Predictable E & O pairing and Case 2 be Incompletely Predictable P & C pairing. Furthermore, let Case 1 and Case 2 be independent of each other. Then for any given x value, there exist grand total number of Dimensions [Complexity] such that it exactly equal to either two combined subtotal number of Dimensions [Complexity] to precisely represent E & O in Case 1, or combined subtotal number of Dimensions [Complexity] to precisely represent P & C & Number '1' in Case 2.

Proof. N is directly constituted from either combined E & O in Case 1 or combined P & C & Number '1' in Case 2 – Number '1' is neither P nor C. Correctly designated infinitely many CFS of Dimensions used to represent combined E & O in Case 1 and combined P & C & Number '1' in Case 2 must also directly and proportionately be representative of relevant N arising from combined subtotal of E & O in Case 1 and from combined subtotal of P & C & Number '1' in Case 2. *The proof is now complete for Proposition 4.1□.*

Proposition 4.2. Let Case 1 be Completely Predictable E & O pairing and Case 2 be Incompletely Predictable P & C pairing. Furthermore, let Case 1 and Case 2 be independent of each other. Part I: For any given x value apart from x = 1 value in Case 1 and x = 1, 2, and 3 values in Case 2; Dimension $(2x - N)^1$ [Complexity] representations of all Completely Predictable E & O in Case 1 and all Incompletely Predictable P & C & Number '1' in Case

2 are such that they are given by N = 4 in Case 1 and by N ≥ 7 in Case 2. Part II: Odd P obeys 'Plus-Minus Composite Gap 2 Number Alternating Law' for prime gaps ≥ 4 and 'Plus Composite Gap 2 Number Continuous Law' for prime gap = 2.

Proof. Apart from first Dimension $(2x - 2)^1$ representation in E & O pairing in Case 1 and first three Dimension $(2x - 2)^1$, Dimension $(2x - 4)^1$ and Dimension $(2x - 5)^1$ representations in P & C pairing in Case 2; possible N value in Dimension $(2x - N)^1$ representation has been shown to be (constantly) maximal 4 for Case 1 and (variably) minimal 7 for Case 2. For Case 2, we again note Dimension $(2x - 2)^1$ to (validly) represent Number '1' which is neither P nor C. These nominated Dimensions simply represent possible (constant) baseline "2x - 4" Grand-Total Gaps as per Proposition 3.7 for Case 1 & (variable) baseline "2x 7" Grand-Total Gaps as per Proposition 3.8 for

189

Case 2. Note that all CFS of Dimensions that can be used to precisely represent combined E & O in Case 1 will persistently consist of same [solitary] Dimension $(2x - 4)^1$ after first Dimension $(2x - 2)^1$. Perpetual repeated deviation of N values away from $N = 7$ (minimum) in Case 2 is simply representing infinite magnitude of P & C. *The proof is now complete for Part I of Proposition 4.2□.*

Derived Dimensions will comply with Incompletely Predictable property as explained using P '61'. At Position $x = 61$ equating to $P_{18} = 61$, it is represented by CFS Dimensions $(2x - 7)^1$, $(2x - 7)^1$, $(2x - 8)^1$, $(2x - 9)^1$, $(2x - 10)^1$ & $(2x - 11)^1$ (with both prime gap & CFS cardinality $= 6$). This representation indicates an 'unknown but correct" P with prime gap $= 6$ when we intentionally conceal full information '61' $= 31^{st}$ O $= 18^{th}$ P with prime gap $= 6$. But to arrive at this representation requires calculations of all preceding CFS Dimensions thus manifesting hallmark Incompletely Predictable property of CFS Dimensions.

Overall sum total of individual CFS Dimensions required to represent every P is infinite in magnitude as $|$all P$| = \aleph_0$. Standalone Dimensions YY groupings [representing signals for "higher order" non-twin P appearances] &/or as front Dimensions YY (sub)groupings [which by itself is fully representative of twin P as Dimensions YYYY appearances] need to recur on an indefinite basis. Then twin P and "higher order" cousin P, sexy P, etc should aesthetically all be infinite in magnitude because (respectively) they regularly and universally arise as part of Dimension YYYY and Dimension YY appearances. An isolated P is defined as a P such that neither P - 2 nor P + 2 is P. In other words, isolated P is not part of a twin P pair. Example 23 is an isolated P since 21 and 25 are both C. Then repeated inevitable presence of Dimension YY grouping is nothing more than indicating repeated occurrences of isolated P. This constitutes another view on Dimension YY.

CIS of Gap 1 Composite Numbers are fully associated with non-twin P as they eternally occur in between any two consecutive non-twin P. CIS of Gap 2 Composite Numbers are (i) fully associated with twin P as they are eternally present in between any twin P pair, and (ii) partially associated with non-twin P as they are eternally present alternatingly or intermittently in between any two consecutive non-twin P. Then (i) Gap 1 Composite Numbers do not have valid representation by E prime gap = 2, and (ii) Gap 2 Composite Numbers have valid representations by all E prime gaps = ["consistently" only for] 2, ["inconsistently" for each of] 4, 6, 8, 10,.... This is an alternative view on P from perspective of CFS composite gaps [instead of CIS prime gaps] with intrinsic patterns having *alternating presence* and *absence* of Gap 2 Composite Numbers associated with every CFS Dimensions representations of P with prime gaps ≥ 4, viz. 'Plus-Minus Gap 2 Composite Number Alternating Law'. CFS Dimensions representations of Twin P are always associated with Gap 2 Composite Numbers, viz. 'Plus Gap 2 Composite Number Continuous Law'.

Examples for both Laws: A twin P (prime gap = 2) in its unique CFS Dimensions format always has Gap 2 Composite Numbers in a [constant] pattern. A cousin P (prime gap = 4) in its unique CFS Dimensions format always has two Gap 1 Composite Numbers & then one Gap 2 Composite Number [combined] pattern *alternating* with three consecutive Gap 1 Composite Numbers [non-combined] pattern. From this simple observation alone, we deduce we can generate an infinite magnitude of C from each composite gaps 1 & 2. Gap 2 Composite Numbers *alternating* pattern behavior in cousin P will not hold true unless twin P & all other non-cousin P are infinite in magnitude and integratedly supplying essential "driving mechanism" to eternally sustain this Gap 2 Composite Numbers *alternating* pattern behavior in cousin P. Thus we establish twin P and cousin P in their CFS Dimensions formats are CIS intertwined together when depicted using C with composite gaps = 1 & 2 with each supplying their own peculiar (infinite) share of associated Gap 2 Composite Numbers [thus contributing to overall pool of Gap 2 Composite Numbers].

An inevitable statement in relation to "Gap 2 Composite Numbers pool contribution" based on above reasoning: At the bare minimum, *either* twin P *or* at least one of non-twin P must be infinite in magnitude. An inevitable impression: All generated subsets of P from 'small gaps' [of 2 & 4] and 'large gaps' [of ≥ 6] alike should each be CIS thus allowing true uniformity in P distribution. Again we see in Table 2

191

above depicting P-C data for x = 2 to 64 that, for instance, P with prime gap = 6 must also persistently have this 'last place' Gap 2 Composite Numbers intermittently appearing in certain rhythmic *alternating* patterns, thus complying with Plus-Minus Gap 2 Composite Number Alternating Law. This CFS Dimensions representation for P with prime gaps = 6 will again generate their infinite share of associated Gap 2 Composite Numbers to contribute to this pool. The presence of this last-place Gap 2 Composite Numbers in various alternating pattern in appearances & nonappearances must *self-generatingly* be similarly extended in a mathematically consistent fashion *ad infinitum* to all other remaining infinite number of prime gaps [which were not discussed in details above]. *The proof is now complete for Part II of Proposition 4.2*□.

5 Rigorous Proofs now named as Polignac's and Twin prime hypotheses

The proofs on lemmas and propositions from previous section supply all necessary evidences to fully support Theorem Polignac-Twin prime I to IV below thus depicting proofs for Polignac's and Twin prime conjectures in a rigorous manner. Gap 1 Composite Numbers do not have valid representation by E prime gap = 2, and Gap 2 Composite Numbers have valid representations by all E prime gaps = ["consistently" only for] 2, ["inconsistently" for each of] 4, 6, 8, 10,.... Plus-Minus Gap 2 Composite Number Alternating Law confirms that Gap 2 Composite Numbers present in each P with prime gaps ≥ 4 situation must appear as some sort of "rhythmic patterns of alternating presence and absence" for Gap 2 Composite Numbers. Twin P with prime gap = 2 obeying Plus Gap 2 Composite Number Continuous Law can be understood as special situation of "(non-)rhythmic patterns with continual presence" for relevant Gap 2 Composite Numbers.

In 1849 when French mathematician Alphonse de Polignac (1826 - 1863) was admitted to Polytechnique, he made what is known as Polignac's conjecture which relates complete set of odd P to all E prime gaps. Twin prime conjecture, which relates twin prime numbers to prime gap = 2, is nothing more than a subset of Polignac's conjecture.

Theorem Polignac-Twin prime I. Incompletely Predictable prime numbers Pn = 2, 3, 5, 7, 11, ..., ∞ or composite numbers Cn = 4, 6, 8, 9, 10, ..., ∞ are CIS with overall actual location [but not actual positions] of all prime or composite numbers accurately represented by complex algorithm involving prime gaps GPi viz. $P_{n+1} = 2$

$$\sum_{+\ i=1}^{n} GPi$$

or involving composite gaps GCi viz. $C_{n+1} = 4 + \sum_{i=1}^{n} GCi$ whereby prime & composite numbers are symbolically represented here with aid of 'n' notation instead of usual 'i' notation; and i & n = 1, 2, 3, 4, 5, ..., ∞. Number '2' in first algorithm represents P1, the very first (and only even) P. Number '4' in second algorithm represent C1, the very first (and even) C.

Proof. We treat above algorithms as unique mathematical objects looking for key intrinsic properties and behaviors. Each P or C is assigned a unique prime or composite gap. Absolute number of P or C and (thus) prime or composite gaps are infinite in magnitude. As original formulae containing all P or C by themselves (viz. without supplying prime or composite gaps as "input information" to generate P or C as "output complexity"), these algorithms intrinsically incorporate overall actual location [but not actual positions] of all P or C. *The proof is now complete for Theorem Polignac-Twin prime I*□.

Theorem Polignac-Twin prime II. Set of prime gaps G_{Pi} = 2, 4, 6, 8, 10, ..., ∞ is infinite in magnitude whereby these prime gaps accurately and completely represented by Dimensions $(2x - 7)^1$, $(2x - 8)^1$, $(2x - 9)^1$, ..., $(2x - \infty)^1$ must satisfy Information-Complexity conservation in a consistent manner.

Proof. Part I of Proposition 4.2 proved all P are represented by Dimension $(2x - N)^1$ with N ≥ 7 for any given x value (except for x = 2 & 3 values). Note that although x = 1 is neither P nor C, it is validly represented by Dimension $(2x - 2)^1$. If each P is endowed with a specific prime gap value, then each such prime gap must [via logical mathematical deduction] be represented by Dimension $(2x - N)^1$. We advocate this nominated method of prime gap representation using Dimensions be [purportedly] the only way to achieve Information-Complexity conservation. The preceding

193

mathematical statements are correct as there is a unique prime gap value associated with each P. Proposition 5.1 below based on principles from Set theory provides further supporting materials that prime gaps are infinite in magnitude. *The proof is now complete for Theorem Polignac-Twin prime II□.*

Theorem Polignac-Twin prime III. To maintain Dimensional analysis (DA) homogeneity, those Dimensions $(2x - N)^1$ from Theorem Polignac-Twin prime II must contain eternal repetitions of well-ordered sets constituted by Dimensions $(2x - 7)^1$, $(2x - 8)^1$, $(2x\ 9)^1$, $(2x - 10)^1$, $(2x - 11)^1$, ..., $(2x - \infty)^1$.

Proof. This Theorem is stated in greater details as "To maintain DA homogeneity, those aforementioned [endowed with exponent 1] Dimensions $(2x - N)^1$ from Theorem Polignac-Twin prime II must repeat themselves indefinitely in following specific combinations – (i) Dimension $(2x - 7)^1$ only appearing as twin [two-times-in-a-row] and quadruplet [fourtimes-in-a-row] sequences, and (ii) Dimensions $(2x - 8)^1$, $(2x - 9)^1$, $(2x - 10)^1$, $(2x - 11)^1$,..., $(2x - \infty)^1$ appearing as progressive groupings of E 2, 4, 6, 8, 10,..., ∞." To accommodate the only even P '2', exceptions to this DA homogeneity compliance will expectedly occur right at beginning of P sequence – (i) one-off appearance of Dimensions $(2x - 2)^1$, $(2x - 4)^1$ and $(2x - 5)^1$ and (ii) one-off appearance of Dimension $(2x - 7)^1$ as a quintuplet [five-times-ina-row] sequence which is equivalent to (eternal) non-appearance of Dimension $(2x - 6)^1$ at x = 4. [We again note Dimension $(2x - 2)^1$ validly represent Number '1' which is neither P nor C.] These sequentially arranged sets are CFS whereby from x = 11 onwards, each set always commence initially as 'baseline' Dimension $(2x - 7)^1$ at x = O values and always end with its last Dimension at x = E values. Each set also have varying cardinality with values derived from all E; and correctly combined sets always intrinsically generate two infinite sets of P and, by default, C in an integrated manner. Our Theorem Polignac-Twin prime III simply represent a mathematical summary derived from Section 3 & 4 of all expressed characteristics of Dimension $(2x - N)^1$ when used to represent P with intrinsic display of DA homogeneity. See Proposition 5.2 below for further details on DA aspect. *The proof is now complete for Theorem Polignac-Twin prime III□.*

194

Theorem Polignac-Twin prime IV. Aspect 1. The "quantitive" aspect to existence of both prime gaps and their associated prime numbers as sets of infinite magnitude will be shown to be correct by utilizing principles from Set theory. Aspect 2. The "qualitative" aspect to existence of both prime gaps and their associated prime numbers as sets of infinite magnitude will be shown to be correct by 'Plus-Minus Gap 2 Composite Number Alternating Law' and 'Plus Gap 2 Composite Number Continuous Law'.

Proof. Required concepts from Set theory involve cardinality of a set with its 'wellordering principle' application. Supporting materials for these concepts based on 'pigeonhole principle' in relation to Aspect 1 are outlined in Proposition 5.1 below. 'Plus-Minus Gap 2 Composite Number Alternating Law' is applicable to all E prime gaps [apart from first E prime gap = 2 for twin primes]. The prime gap = 2 situation will obey 'Plus Gap 2 Composite Number Continuous Law'. These Laws are essentially Laws of Continuity inferring underlying intrinsic driving mechanisms that enables infinity magnitude association for both prime gaps & prime numbers to co-exist. By the same token, these Laws have the important implication that they must be applicable to those relevant prime gaps on a perpetual time scale. Supporting materials in relation to Aspect 2 are found in Proposition 4.2 above. *The proof is now complete for Theorem Polignac-Twin prime IV*□.

We note two mutually inclusive conditions: Condition 1. Presence of all Dimensions that repeat themselves on an indefinite basis and with exponent of '1' will give rise to complete sets of P & C ["DA-wise one & only one mathematical possibility argument" associated with inevitable *de novo* DA homogeneity], and Condition 2. Presence of any Dimension(s) that do not repeat itself (themselves) on an indefinite basis or with exponent other than '1' will give rise to incomplete set of P & C or incorrect set of non-P & non-C ["DA-wise mathematical impossibility argument" associated with inevitable *de novo* DA non-homogeneity]. When met, these two conditions will fully support the point that CFS Dimensions representations of P & C [with respective prime & composite gaps] are totally accurate. Condition 1 reflect proof

195

from Theorem Polignac-Twin prime III above as all P & C are associated with DA homogeneity when their Dimensions are endowed with exponent of '1'. Condition 2 invoke corollary on inevitable appearance of incomplete P or C or non-P or non-C [associated with DA non-homogeneity] being tightly incorporated into this mathematical framework. See Propositions 5.1 and 5.2, and Corollary 5.3 below for supporting materials on DA homogeneity & non-homogeneity.

We analyze P (& C) in terms of (i) measurements based on cardinality of CIS and (ii) pigeonhole principle which states that if n items are put into m containers, with n>m, then at least one container must contain more than one item. We note that ordinality of all infinite P (& C) is "fixed" implying that each one of the infinite well-ordered Dimension sets conforming to CFS type as constituted by Dimensions $(2x - 7)^1$, $(2x - 8)^1$, $(2x - 9)^1$, $(2x - 10)^1$, $(2x - 11)^1$, ..., $(2x - \infty)^1$ on respective gaps for P (& C) must also be "fixed".

Proposition 5.1. "Even number prime gaps are infinite in magnitude with each even number prime gap generating odd prime numbers which are again infinite in magnitude" is supported by principles from Set theory and two Laws based on Gap 2 Composite Number.

Proof. We validly exclude even P '2' here. Let (i) cardinality $T = \aleph_0$ for Set all odd P derived from E prime gaps 2, 4, 6,..., ∞, (ii) cardinality $T_2 = \aleph_0$ for Subset odd P derived from E prime gap 2, cardinality $T_4 = \aleph_0$ for Subset odd P derived from E prime gap 4, cardinality $T_6 = \aleph_0$ for Subset odd P derived from E prime gap 6, etc. Paradoxically $T = T_2 + T_4 + T_6 +... + T_\infty$ equation is valid despite $T = T_2 = T_4 = T_6 =... = T_\infty$ [with well-ordering principle "stating that every non-empty set of positive integers contains a least element" fulfilled by each (sub)set]. But if Subset odd P derived from one or more E prime gap(s) are finite in magnitude, this will breach $\aleph_0$ 'uniformity' resulting in (i) DA non-homogeneity and (ii) inequality $T > T_2 + T_4 + T_6 +... + T_\infty$. In language of pigeonhole principle "stating that if n items are put into m containers with n>m, then at least one container must contain more than one item", residual odd P (still CIS in magnitude) not accounted for by CFS type E prime gap(s) will have to be [incorrectly]

196

contained in one (or more) of composite gap(s). These arguments using cardinality constitute proof that (i) E prime gaps and (ii) odd P generated from each E prime gap, must all be CIS. *The proof [on "quantitative" aspect] is now complete for Proposition 5.1□.*

Complete set of P is represented by Dimensions $(2x - N)^1$. Table 2 & Figure 1 on PC finite scale mathematical landscape depict perpetual repeating features used in "qualitative" statements supporting (i) Plus-Minus Gap 2 Composite Number Alternating Law (stated as C with composite gaps = 2 present in each of P with prime gaps ≥ 4 situation must be observed to appear as some sort of rhythmic patterns of alternating presence and absence of this type of C), and (ii) Plus Gap 2 Composite Number Continuous Law (stated as C with composite gaps = 2 continual appearances in each of (twin) P with prime gap = 2 situation). Plus-Minus Gap 2 Composite Number Alternating Law has intrinsic mechanism to automatically generate all prime gaps ≥ 4 in a mathematically consistent *ad infinitum* manner. Plus Gap 2 Composite Number Continuous Law has built-in intrinsic mechanism to further generate prime gap = 2 appearances in a mathematically consistent *ad infinitum* manner. *The proof [on "qualitative" aspect] is now complete for Proposition 5.1□.*

Proposition 5.2. The presence of Dimensional analysis homogeneity will always result in correct and complete set of prime (and composite) numbers.

Proof. DA homogeneity is completely dependent on all Dimensions being consistently endowed with exponent '1'. As all P (& C) are "fixed", we deduce from Figure 1 & Table 2 that there is one (& only one) way to represent Information-Complexity conservation using our defined Dimensions. Thus, there is one (& only one) way to depict all P (& C) using these Dimensions in a self-consistent manner and this is achieved with the one (& only one) DA homogeneity possibility. *The proof is now complete for Proposition 5.2□.*

Corollary 5.3. The presence of Dimensional analysis non-homogeneity will always result in incorrect and/or incomplete set of prime (and composite) numbers.

Proof. For optimal clarity, we endow all Dimensions with exponent '1' depicted as $(2x - 7)^1$, $(2x - 8)^1$, $(2x - 9)^1$, $(2x - 10)^1$, $(2x - 11)^1$,..., $(2x - \infty)^1$. Proposition 5.2 equates

197

DA homogeneity with correct & complete set of P (& C). There are "more than one" DA possibilities when, for instance, a particular [first] term from $(2x - 7)^0$, $(2x - 8)^1$, $(2x - 9)^1$,..., $(2x - \infty)^1$ "terminates" prematurely and does not perpetually repeat [with loss of continuity]. There are intuitively two 'broad' DA possibilities here; namely, (one) DA homogeneity possibility and (one) DA non-homogeneity possibility – Dimension $(2x - 7)^0$ [= 1] with its exponent arbitrarily set as '0' against-all-trend in this case. Thus Dimension $(2x - 7)^1$ that stop recurring at some point in P (or C) sequence may cause well-ordered CFS sets from progressive groupings of [E] 2, 4, 6, 8, 10,..., ∞ for Dimensions $(2x - 8)^1$, $(2x - 9)^1$, $(2x - 10)^1$, $(2x - 11)^1$,..., $(2x - \infty)^1$ to stop existing (and ultimately for sequential P (or C) to stop appearing) at that point with ensuing outcome that P (or C) may overall be incorrectly finite or incomplete in magnitude. Finally also manifesting DA non-homogeneity, a Dimension endowed with fractional exponent values other than '1' such as '$\frac{2}{5}$' or '$\frac{3}{5}$' will result in non-P (or non-C) [fractional] numbers. *The proof is now complete for Corollary 5.3*□.

Each [fixed] finite scale mathematical landscape "page" as part of [fixed] infinite scale mathematical landscape "pages" for P & C display Chaos [sensitivity to initial conditions viz. positions of subsequent P & C are "sensitive" to positions of initial P & C] and Fractals [manifesting fractal dimensions with self-similarity viz. those aforementioned Dimensions for P & C are always present, albeit in non-identical manner, for all ranges of $x \geq 2$]. Advocated in another manner, Chaos and Fractals phenomena of those Dimensions for P & C are always present signifying accurate composition of P & C in different [predetermined] finite scale mathematical landscape "(snapshot) pages" for P & C that are self-similar but never identical – and there are an infinite number of these finite scale mathematical landscape "(snapshot) pages". The crucial mathematical step in representing all P (& C) and prime (& composite) gaps with "Dimensions" based on Information-Complexity conservation allows us to obtain the two Laws based on Gap 2 Composite Numbers and perform DA on these entities. The 'strong' principle argument is DA homogeneity equates to complete set of P (& C) whereas DA non-homogeneity does not equate to complete set of P (& C). We could also advocate for a 'weak' principle argument supporting DA homogeneity for P (& C) in that nature should not "favor" any particular Dimension(s) to terminate and therefore DA non-homogeneity does not, and cannot, exist for P (& C). Abiding to our advocated convention that 'conjecture' be termed 'hypothesis' once proven; we now call Polignac's & Twin prime conjectures as Polignac's & Twin prime hypotheses.

6 Conclusions

Harnassed property: CIS of [Completely Predictable] natural numbers 1, 2, 3, 4, 5, 6, 7,... having CIS of [Completely Predictable] natural gaps 1, 1, 1, 1, 1, 1,... are constituted by three dependent sets of numbers: (i) CIS of [Incompletely Predictable] odd prime numbers 3, 5, 7, 11, 13, 17,... having CIS of [Incompletely Predictable] prime gaps 2, 2, 4, 2, 4,... plus CFS of solitary [Incompletely Predictable] even prime number 2 having CFS of [Incompletely Predictable] prime gap 1 (ii) CIS of [Incompletely Predictable] even and odd composite numbers 4, 6, 8, 9, 10, 12,... having CIS of [Incompletely Predictable] composite gaps 2, 2, 1, 1, 2, 2,.... and (iii) CFS of solitary odd number '1' [neither prime nor composite]. Treated as Incompletely Predictable problems endowed with "meta-properties", we gave relatively elementary proofs on Polignac's and Twin prime conjectures based on this harnessed property by performing Dimensional analysis on (sub)sets and "Dimensions" of prime and composite numbers, and obtaining 'Plus-Minus Gap 2 Composite Number Alternating Law' and 'Plus Gap 2 Composite Number Continuous Law'.

Prime number theorem describes asymptotic distribution of prime numbers among positive integers by formalizing intuitive idea that prime numbers become less common as they become larger through precisely quantifying rate at which this occurs using probability. Nontrivial zeros [from 'Axes intercept relationship interface' relevant to Riemann hypothesis] and prime numbers [from 'Numerical relationship interface' relevant to prime number theorem] are Incompletely Predictable entities and numbers. Deep-seated connections exist between Riemann hypothesis and prime number theorem (which is fully delineated by prime counting function [denoted here with

$\pi(x)$]). Solving Incompletely Predictable problem Riemann hypothesis is instrumental in proving efficacy of techniques that estimate $\pi(x)$ efficiently. This should now confirm "best possible" bound for error ("smallest possible" error) of prime number theorem.

In mathematics, logarithmic integral function or integral logarithm li(x) is a special function. Relevant to problems of physics and with number theoretic significance, it occurs in prime number theorem as an estimate of $\pi(x)$ whereby the form of this special function is defined so that li(2) = 0; viz. li(x) $= \int_2^x \frac{du}{\ln u} =$ li(x) - li(2). There are less accurate ways of estimating $\pi(x)$ such as conjectured by Gauss and Legendre at end of 18th century. This $\pi(x)$ is approximately x/ln x in the sense sense $\lim_{x\to\infty} \left(\frac{\pi(x)}{nx/\ln x} \right) = 1$. Skewes' number is any of several extremely large numbers used by South African mathematician Stanley Skewes as upper bounds for smallest natural number x for which li(x)<$\pi(x)$. These bounds have since been improved by others: there is a crossing near $e^{727.95133}$ but it is not known whether this is the smallest. John Edensor Littlewood, who was Skewes' research supervisor, proved in 1914[4] that there is such a [first] number; and found that sign of difference $\pi(x)$ - li(x) changes infinitely often. This refute all prior numerical evidence that seem to suggest li(x) was always more than $\pi(x)$. The key point is [100% accurate] $\pi(x)$ mathematical tool being "wrapped around" by [less-than-100% accurate] approximate mathematical tool li(x) infinitely often via this 'sign of difference' changes meant that li(x) is the most efficient approximate mathematical tool. Contrast this with "crude" x/lnx approximate mathematical tool where values obtained diverge away from $\pi(x)$ at increasingly greater rate when larger range of prime numbers are studied.

Table 3 Even-Odd mathematical (tabulated) landscape using data obtained for x = 1 to 64. Legend: E=even, O=odd, Y=Dimension 2x-4.

x	E_i or O_i, Gaps	ΣEO_x - Gaps	Dimension

1	O1, 2	0	2x-2
2	E1, 2	0	Y
3	O2, 2	2	Y
4	E2, 2	4	Y
5	O3, 2	6	Y
6	E3, 2	8	Y
7	O4, 2	10	Y
8	E4, 2	12	Y
9	O5, 2	14	Y
10	E5, 2	16	Y
11	O6, 2	18	Y
12	E6, 2	20	Y
13	O7, 2	22	Y
14	E7, 2	24	Y
15	O8, 2	26	Y
16	E8, 2	28	Y
17	O9, 2	30	Y
18	E9, 2	32	Y
19	O10, 2	34	Y
20	E10, 2	36	Y
21	O11, 2	38	Y
22	E11, 2	40	Y
23	O12, 2	42	Y
24	E12, 2	44	Y
25	O13, 2	46	Y
26	E13, 2	48	Y
27	O14, 2	50	Y

28	E14, 2	52	Y
29	O15, 2	54	Y
30	E15, 2	56	Y
31	O16, 2	58	Y
32	E16, 2	60	Y
33	O17, 2	62	Y
34	O17, 2	64	Y
35	O17, 2	66	Y
36	O17, 2	68	Y
37	O17, 2	70	Y
38	O17, 2	72	Y
39	O17, 2	74	Y
40	O17, 2	76	Y
41	O17, 2	78	Y
42	O17, 2	80	Y
43	O17, 2	82	Y
44	O17, 2	84	Y
45	O17, 2	86	Y
46	O17, 2	88	Y
47	O17, 2	90	Y
48	O17, 2	92	Y
49	O17, 2	94	Y
50	O17, 2	96	Y
51	O17, 2	98	Y
52	O17, 2	100	Y
53	O17, 2	102	Y
54	O17, 2	104	Y

55	O17, 2	106	Y
56	O17, 2	108	Y
57	O17, 2	110	Y
58	O17, 2	112	Y
59	O17, 2	114	Y
60	O17, 2	116	Y
61	O17, 2	118	Y
62	O17, 2	120	Y
63	O17, 2	122	Y
64	O17, 2	124	Y

Appendix I: Tabulated and graphical data on Even-Odd mathematical landscape

We tabulate (in Table 3) and graph (in Figure 2) [Completely Predictable] E-O mathematical landscape for x = 1 to 64. Involved Dimensions are 2x - 2 & 2x - 4 with Y denoting Dimension 2x - 4 for visual clarity. This mathematical landscape of Dimension 2x - 4 (except for first and only Dimension 2x - 2) will intrinsically incorporate E & O in an integrated manner. Except for first O, all Completely Predictable E & O and all their associated gaps are represented by countable finite set of [single] Dimension 2x - 4. Dimensions 2x - 2 & 2x - 4 are symbolically represented by -2 & -4 with 2x - 4 displayed as 'baseline' Dimension whereby Dimension trend (Cumulative Sum Gaps) must reset itself onto this (Grand-Total Gaps) 'baseline' Dimension after initial Dimension 2x - 2 on a permanent basis. Graphical appearances of Dimensions symbolically represented by two negative integers are Completely Predictable with both Even-$\pi(x)$ and Odd-$\pi(x)$ becoming larger at a constant rate. There is a complete absence of Chaos and Fractals phenomena.

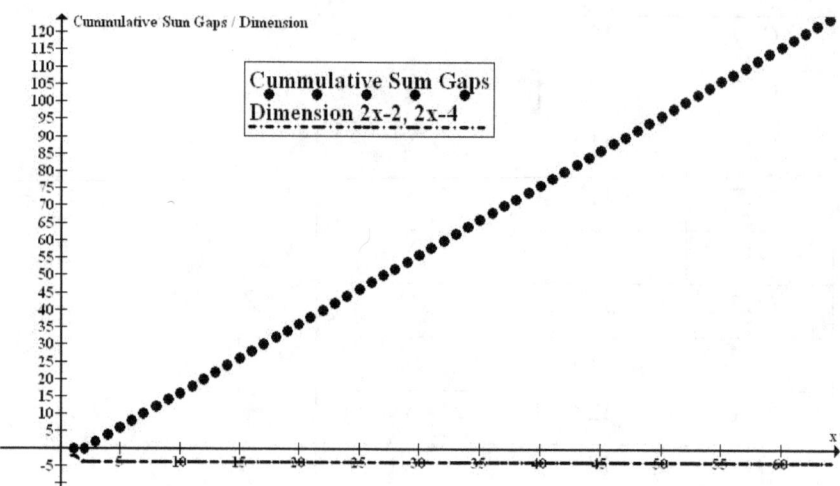

Fig. 2 Even-Odd mathematical (graphed) landscape using data obtained for x = 1 to 64.

Definitive derivation of data in Table 3 is illustrated by two examples for position x = 31 & 32. For i & x ∈ 1, 2, 3, ..., ∞; ΣEO_x-Gap = ΣEO_{x-1}-Gap + Gap value at E_{i-1} or Gap value at O_{i-1} whereby (i) E_i or O_i at position x is determined by whether relevant x value belongs to E or O, and (ii) both ΣEO_1-Gap and ΣEO_2-Gap = 0. Example, for position x = 31: 31 is O (O16). Our desired Gap value at O15 = 2. Thus ΣEO_{31}-Gap (58) = ΣEO_{30}-Gap (56) + Gap value at O15 (2). Example, for position x = 32: 32 is E (E16). Our desired Gap value at E15 = 2. Thus ΣEO_{32}-Gap (60) = ΣEO_{31}-Gap (58) + Gap value at E15 (2).

Acknowledgements I am indebted to Mr. Rodney Williams and Mr. Tony O'Hagan for reviewing this paper (dedicated to my daughter Jelena born 13 weeks early on May 14, 2012).

References

1. Furstenberg, H. (1955). On the infinitude of primes. *Amer. Math. Monthly, 62*, (5) 353, http://dx.doi.org/10.2307/2307043

2. Saidak, F. (2006), A New Proof of Euclid's theorem, *Amer. Math. Monthly, 113*, (10) 937, http://dx.doi.org/10.2307/27642094

3. Zhang, Y. (2014), Bounded gaps between primes, *Ann. Math. 179*(3) (2014) 1121 – 1174, http://dx.doi.org/10.4007/annals.2014.179.3.7

4. Littlewood, J. E. (1914), Sur la distribution des nombres premiers. Comptes Rendus de l'Acad. Sci. Paris, 158, 1869 – 1872

ABOUT THE AUTHOR

My novel Hybrid integer sequence A228186 was published in The On-Line Encyclopedia of Integer Sequences on August 15, 2013. From 2016 to 2019, I carry out mathematical research with published papers in Number theory on Riemann hypothesis, Polignac's and Twin prime conjectures.

I live in Brisbane, Australia with my wife and five children. I possess Australian Medical degree MBBS (1989), General Practice FRACGP (1994), Primary Anesthesia Fellowship Examination (2009) and Opioid Replacement license (2017). My work experiences have involve the specialty area of Anesthesia, Intensive Care, Pain Medicine, Medicinal Cannibis and Addiction Medicine. My medical publication as primary author with Professor Bruce A. Pussell (secondary author) include Supramaximal elevation in B-type natriuretic peptide and its N-terminal fragment levels in anephric patients with heart failure: a case series. Journal of medical case reports, 2012.